ESTROGEN EFFECTS ON TRAUMATIC BRAIN INJURY

ESTROGEN EFFECTS ON TRAUMATIC BRAIN INJURY

Mechanisms of Neuroprotection and Repair

Edited by

Kelli A. Duncan

AMSTERDAM • BOSTON • HEIDELBERG • LONDON
NEW YORK • OXFORD • PARIS • SAN DIEGO
SAN FRANCISCO • SINGAPORE • SYDNEY • TOKYO
Academic Press is an imprint of Elsevier

Academic Press is an imprint of Elsevier
32 Jamestown Road, London NW1 7BY, UK
525 B Street, Suite 1800, San Diego, CA 92101-4495, USA
225 Wyman Street, Waltham, MA 02451, USA
The Boulevard, Langford Lane, Kidlington, Oxford OX5 1GB, UK

ISBN: 978-0-12-801479-0

British Library Cataloguing-in-Publication Data
A catalogue record for this book is available from the British Library

Library of Congress Cataloging-in-Publication Data
A catalog record for this book is available from the Library of Congress

For information on all Academic Press publications
visit our website at http://store.elsevier.com/

Typeset by MPS Limited, Chennai, India
www.adi-mps.com

Printed and bound in the United States of America

Contents

About the Editor

Kelli A. Duncan is a professor of biology and neuroscience and behavior at Vassar College. She provides the following insight into her career and research.

My lab studies the role of steroid hormones, specifically estrogens and androgens following traumatic brain injury, working exclusively with undergraduates. As I look back at my career, I notice that I have always been interested in steroid hormones in some way or another. My first experience with steroid hormones began on a bus stop during my senior year at the University of Georgia when I was randomly approached to apply for a summer research program. I decided to apply, but I needed to find a research mentor. As an undergraduate, I had taken a number of classes in psychology and biology, but was always planning on becoming a veterinarian. Thus, when looking for a lab to work in, I wanted to find a lab where I could have the opportunity to work with animals or study behavior. I landed in a lab doing work in mosquitoes examining ecdysteroid synthesis in female mosquitoes following a blood meal. Ecdysteroids are the insect version of steroid hormones. I soon realized that I really enjoyed research and wanted to remain in the field, so instead of going to veterinary school, I decided to go to graduate school and focus on sex differences in behavior. I was given a list of prospective schools that focused on behavioral neuroscience. One set of schools belonged to a consortium, The Center for Behavioral Neuroscience (CBN). The CBN, a National Science Foundation Science and Technology Center, was a consortium of seven metro Atlanta colleges and universities: Georgia State University, Emory University, Georgia Tech, Clark Atlanta University, Morehouse College, Morehouse School of Medicine, and Spelman College. More than 150 neuroscientists led the research program with the goal of understanding the basic neurobiology of social behaviors. My graduate career within the CBN expanded my interest in behavioral neuroscience, specifically in terms of sex differences. My doctoral work examined sex differences in brain morphology and singing behavior in zebra finches and focused on the role of estrogens and estrogen receptor coactivators. Wanting to continue in the estrogen world, I took a postdoctoral position examining the role of estrogens following traumatic brain injury in zebra finches and sex differences following injury. This interaction between sex differences and estrogens continues to fuel my work today, including the expansion of my research into steroid mediation of neuroinflammation.

List of Contributors

Marco Ávila-Rodriguez Departamento de Nutrición y Bioquímica, Facultad de Ciencias, Pontificia Universidad Javeriana, Bogotá, Colombia

Taura L. Barr Center for Basic and Translational Stroke Research, West Virginia University Health Sciences Center, Morgantown, West Virginia, USA; School of Nursing, West Virginia University, Morgantown, West Virginia, USA

George E. Barreto Departamento de Nutrición y Bioquímica, Facultad de Ciencias, Pontificia Universidad Javeriana, Bogotá, Colombia

Suzanne R. Burstein Weill Cornell Medical College, New York, New York, USA

Francisco Capani Departamento de Biología, Universidad Argentina John F Kennedy, Buenos Aires, Argentina; Laboratorio de Citoarquitectura e Injuria Neuronal, Instituto de Investigaciones Cardiológicas "Prof. Dr. Alberto C. Taquini" (ININCA), UBA-CONICET, Buenos Aires, Argentina; Facultad de Psicología, Universidad Católica Argentina, Buenos Aires, Argentina

Laura L. Carruth Neuroscience Institute, Georgia State University, Atlanta, Georgia, USA

Rocío Castilla Laboratorio de Citoarquitectura e Injuria Neuronal, Instituto de Investigaciones Cardiológicas "Prof. Dr. Alberto C. Taquini" (ININCA), UBA-CONICET, Buenos Aires, Argentina

Pascal Coumailleau Neuroendocrine Effects of Endocrine Disruptors, IRSET, INSERM U1085 Université de Rennes, Rennes, France

Nicolas Diotel INSERM U1188, Plateforme CYROI, Université de La Réunion, Saint Denis de La Réunion, France

Kelli A. Duncan Biology and Neuroscience and Behavior, Vassar College, Poughkeepsie, New York, USA

Luis Miguel García-Segura Instituto Cajal, CSIC, Madrid, Spain

Janneth Gonzalez Departamento de Nutrición y Bioquímica, Facultad de Ciencias, Pontificia Universidad Javeriana, Bogotá, Colombia

John P. Hayden Biology, Vassar College, Poughkeepsie, New York, USA

Olivier Kah Neuroendocrine Effects of Endocrine Disruptors, IRSET, INSERM U1085 Université de Rennes, Rennes, France

Rodolfo Kolliker Laboratorio de Citoarquitectura e Injuria Neuronal, Instituto de Investigaciones Cardiológicas "Prof. Dr. Alberto C. Taquini" (ININCA), UBA-CONICET, Buenos Aires, Argentina

Ana Belen Lopez-Rodriguez Departamento de Fisiología Animal (II), Facultad de Biología, Universidad Complutense de Madrid, Spain; Instituto Cajal, CSIC, Madrid, Spain

Carissa J. Mehos Departments of Biology, Psychology and the Center for Behavioral Neuroscience, American University, Washington, DC, USA

Alyssa L. Pedersen Departments of Biology, Psychology and the Center for Behavioral Neuroscience, American University, Washington, DC, USA

Elisabeth Pellegrini Neuroendocrine Effects of Endocrine Disruptors, IRSET, INSERM U1085 Université de Rennes, Rennes, France

Ashley B. Petrone Center for Basic and Translational Stroke Research, West Virginia University Health Sciences Center, Morgantown, West Virginia, USA; Center for Neuroscience; School of Nursing, West Virginia University, Morgantown, West Virginia, USA

Miranda N. Reed Center for Basic and Translational Stroke Research, West Virginia University Health Sciences Center, Morgantown, West Virginia, USA; Center for Neuroscience, West Virginia University, Morgantown, West Virginia, USA; Department of Psychology, West Virginia University, Morgantown, West Virginia, USA; Center for Neuroscience, West Virginia University, Morgantown, West Virginia, USA

Carolyn C. Rudy Department of Psychology, West Virginia University, Morgantown, West Virginia, USA

Colin J. Saldanha Departments of Biology, Psychology and the Center for Behavioral Neuroscience, American University, Washington, DC, USA

Ezequiel Saraceno Laboratorio de Citoarquitectura e Injuria Neuronal, Instituto de Investigaciones Cardiológicas "Prof. Dr. Alberto C. Taquini" (ININCA), UBA-CONICET, Buenos Aires, Argentina

Mahin Shahbazi Neuroscience Institute, Georgia State University, Atlanta, Georgia, USA

James W. Simpkins Center for Basic and Translational Stroke Research, West Virginia University Health Sciences Center, Morgantown, West Virginia, USA; Department of Physiology and Pharmacology, School of Medicine, West Virginia University, Morgantown, West Virginia, USA

Farida Sohrabji Women's Health in Neuroscience Program, Neuroscience and Experimental Therapeutics, Texas A&M Health Science Center, College of Medicine, Bryan, Texas, USA

Nelson E. Vega-Vela Departamento de Nutrición y Bioquímica, Facultad de Ciencias, Pontificia Universidad Javeriana, Bogotá, Colombia

William A. Wiggins Departments of Biology, Psychology and the Center for Behavioral Neuroscience, American University, Washington, DC, USA

About the Contributors

TAURA BARR

My career began as a registered nurse in the Neurovascular Intensive Care Unit (ICU) at the University of Pittsburgh Presbyterian Hospital. In a short period of time at the bedside, I realized that clinical care of stroke and brain-injured patients could be improved. Because of these gaps in clinical care, I became passionately motivated to help my patients realize their full recovery potential starting with the care they received under my shift in the ICU. After critical discussions with a nurse scientist mentor, I recognized that a career in research would help me reach my end goal of decreasing the burdens of brain injury by changing the way brain injuries are studied and then ultimately treated. Within a year of undergraduate graduation, I entered a BSN-Ph.D. program at the University of Pittsburgh and became the first Graduate Partnership Program (GPP) fellow at the university (www.training.nih.gov/programs/gpp). The National Institutes of Health (NIH) GPP program allows graduate students to complete coursework at their home institution, and then complete dissertation work on the intramural NIH campus with NIH mentors and unlimited research resources. As a graduate student of the intramural National Institute of Nursing Research (NINR) program, I was part of a cutting-edge research team that was fully integrated into clinical practice of stroke patients in the National Institute of Neurological Disorders and Stroke (NINDS), Stroke Diagnostics and Therapeutics Section. The mentoring I received and the experience of developing and implementing a genomics-based biomarker study for ischemic stroke as a graduate student was critical to my development as a clinical translational neuroscientist and my subsequent program of research. After completion of my dissertation, *Biomarkers of Acute Stroke*, I was offered a Nurse Specialist appointment in the intramural NINR program to develop and implement a partnership with the Center for Neuroscience and Regenerative Medicine (CNRM) at the Uniformed Services University (USU), Bethesda, MD. I was part of a team of clinicians and researchers setting the research agenda for traumatic brain injury (TBI) in military patients in the national capital area. Severe TBI and concussion (or mild TBI) affects over 30% of all deployed military personnel and can result in significant long-term morbidity and psychological health issues. The long-term complications of TBI are not just a problem in the military; civilian TBI can also result in posttraumatic stress syndrome (PTSD) and postconcussion syndrome (PCS) in 24–80% of cases. We have yet to define the underlying physiology that puts patients with TBI at risk for these debilitating psychological health issues and I believe the mechanisms begin during the acute injury period. Using my prior clinical knowledge of civilian TBI, my experience as a graduate student in the stroke program, observations with clinical practice at Walter Reed and through collaboration with CNRM investigators I became extremely motivated to study the physiological underpinnings of these conditions post-TBI. Our team designed a study and received funding to partner with a Sleep Medicine physician at an Army Medical Center in Tacoma, WA to study sleep and identify a genomic profile in the peripheral blood of post-deployed military personnel who had suffered a TBI in the theater of war. We wanted to know if a combination of sleep disturbances and a specific genomic expression profile could be used to predict patients who went on to develop PCS and/or PTSD. Unfortunately, when I left the NIH to begin my faculty appointment at West Virginia University I had to transfer the principal investigator role to a colleague in the intramural NINR program. She kept the study going until its completion in Fall 2013. We have recently published some of this work, while the remainder of the data is being analyzed. This work had a tremendous impact on the clinical care of patients at MAMC and our goal is to integrate the findings in other settings, both military and civilian, to raise awareness of sleep disorders in brain-injured patients and identify novel treatment approaches to mitigate the increased risk of psychological health problems. Since this study, I have continued my work on ischemic stroke and TBI with the goal of identifying novel biomarkers that can be used in clinical care for diagnosis, but then taken back to the bench to study their functional relevance. My program of research starts with the patient at the bedside, and all of the variability and clinical issues that come with them; brings those issues back to the bench to identify treatment strategies; and then ends with industry partnerships to commercialize and fully translate our results. Given the paramount important of increasing the speed and efficiency of clinical translation, the changes in our medical system that will come with the Affordable Care Act, and decreasing sources of research funding, the future of brain injury research will depend on innovative partnerships between academia, hospitals, government and industry to tackle the challenges of brain injury recovery together. It will be imperative for future investigators in the field to be well-rounded and have knowledge in best clinical practice and industry initiatives to ensure the work they are doing at the bench does indeed make it to the patient at the bedside.

GEORGE E. BARRETO

I earned my B.Sc. in physical therapy in 2002 and M.Sc. in biochemistry in 2005. A year before finalizing my M.Sc. training, I was already thinking about the next step to pursue in my academic career. I looked over some good Ph.D. programs in Brazil, but none attracted me. As I have had a clinical influence in my entire research career in college, I wanted to study something that could be translated into a clinical approach. I was particularly sensitive to patients with neurological impairment after some types of brain injury. My idea was to have a molecular/cellular basis background of these neurological alterations to broaden my knowledge, and target some possible therapeutic interventions to ameliorate and improve neurological outcomes. With this in mind, I searched for good laboratories. Finally, I was told that the Cajal Institute in Madrid was known for its reputation in the neurosciences field, especially my Ph.D. advisor, Dr. Luis Miguel Garcia-Segura. Since I have always been into neurological sciences and had positive feelings for pursuing such an education abroad, with such great perspectives in brain research, the next step was to find a fellowship that could support me financially. I was awarded a prestigious and competitive Western European Research Committee (WERC)/Federation of European Neuroscience Societies (FENS)/International Brain Research Organization (IBRO) Ph.D. fellowship (2006–2008). I successfully finalized my Ph.D. studies on January 9, 2009 from Universidad Complutense Madrid (Spain). During my Ph.D. studies, I addressed the effects of neurosteroids on reactive glia with aging. My postdoctoral training at Stanford University School of Medicine (2009–2011) focused on addressing the role of astrocyte activation following stroke and enhanced astrocytic functions targeting neuronal protection. My research interests relate to how potential therapeutic strategies may control astrogliosis and therefore improve neuronal survival, focusing particularly on the two-sided role of reactive astrocytes, which is an experimental paradigm helpful in discriminating destructive from protective mechanisms during the normal aging process and after brain injury.

SUZANNE R. BURSTEIN

Currently, I am a graduate student at Weill Cornell Medical College. I became interested in the neuroscience field as an undergraduate student studying biology at Lehigh University, and began working in the neuroendocrinology laboratory of Colin Saldanha, Ph.D. After realizing in my undergraduate courses that there is still so much that remains unknown about the complexities of the human brain, I was excited about a future in neuroscience research. In the Saldanha lab, I studied sex differences in cytokine induction and aromatization after traumatic brain injury in the zebra finch. After graduating, my interest in the role of estrogens in neuroprotection remained and I decided to pursue a Ph.D, where I now study the role of estrogen in mitochondrial dysfunction in neurodegeneration and stroke.

FRANCISCO CAPANI

Since I am a physician, my interest in neuroprotection began early in my career with my graduate training at the School of Medicine of the University of Buenos Aires. I trained during my Ph.D. studies with Prof. Dr. Jorge Pecci Saavedra in a model of perinatal asphyxia in rat using different protocols of experimental hypothermia. After my postdoctoral formation at the Department of Neuroscience, University of San Diego, and at the Department of Neuroscience, Karolinska Institutet, I returned to the School of Medicine, University of Buenos Aires, in 2006, where I obtained a faculty position. Together with some talented students and in collaboration with Dr. Luis Miguel Garcia Segura, we found that estrogen was neuroprotective on the long-term alterations induced by perinatal asphyxia, this experimental therapeutic approach being more specific than the effect of hypothermia. These results still have a strong influence on our current research, developing new therapeutic approaches to find a cure for the devastating consequences of perinatal asphyxia.

LAURA L. CARRUTH

The hormonal regulation of brain and behavior in vertebrates has fascinated generations of scientists, including myself, my interest germinating as an undergraduate student and subsequently influencing my graduate education. My training in endocrinology started as a graduate student at the University of Colorado in Boulder, where I trained in the laboratories of Drs. Richard E. Jones and David O. Norris, studying the endocrine regulation of stress and reproduction in a variety of nonmammalian vertebrates, primarily different species of fish, amphibians and reptiles. What fascinated me and others in the lab was knowing that, while the functions of different steroids had changed multiple times over evolutionary time, their chemical structures were still highly conserved and nearly identical among different vertebrate animals. For my postdoctoral training I moved up the evolutionary ladder, so to speak, to study songbirds with Dr. Arthur P. Arnold at University of California at Los Angeles (CA), whose contributions to the field of behavioral neuroendocrinology are numerous. His research on sex differences in the vertebrate brain and behavior involving hormones and direct genetic mechanisms is highly influential, and in his lab we worked to identify factors involved in the development of the sexually dimorphic neuronal song control circuit. Sexual differentiation of the songbird brain does not appear to completely follow the mechanisms identified for the mammalian brain. After obtaining my current faculty position, I continued on with the hunt to identify a factor or factors that might be involved in this process. Working with my first graduate student, Dr. Kelli Duncan, we came a little closer with the identification of cofactor proteins involved in steroid induced transcription. Parallel to this line of research another graduate student, Dr. Mahin Shahbazi, started a project investigating the role of stress in early development of the avian song system, pulling me back to my early roots in stress research. My lab continues to investigate both avenues of research.

NICOLAS DIOTEL

My interest in the roles of estrogens on adult neurogenesis began during my practical work at the Bachelors Degree level. I became fascinated by the fact that the adult brain of fish exhibits a strong aromatase activity and a high neurogenesis. Thus, I decided to apply for a Ph.D. position at the University of Rennes 1 (France) in the laboratory of a well-known neuroendocrinologist, Dr. Olivier Kah. A few years before I applied, his group was the first to show that aromatase expressing cells were neural progenitors in adult zebrafish. Dr. Kah and my two other co-supervisors, Dr. Elisabeth Pellegrini and Dr. Colette Vaillant, provided me an excellent training and research environment where I could gain strong knowledge about how estrogens and steroids can affect brain physiology, plasticity, and neurogenesis. During my Ph.D. program, we notably demonstrated that the brain of adult zebrafish is a source and a target of steroids and that adult neurogenesis is regulated by the estrogenic environment. Next, I decided to do my postdoctoral training in the laboratory of Pr. Dr. Uwe Straehle, an expert in neurogenesis. In his group at the Karlsruhe Institute of Technology (Germany), I worked on the roles of transcription factors in the maintenance and recruitment of neural stem cells in adult brains under both homeostatic and injury conditions. Currently, I am an assistant professor at the University of La Réunion (France) where I am working on the better understanding of how metabolic diseases such as obesity or diabetes could promote neuroinflammation and neurodegeneration; and how estrogens, steroids and other factors could have beneficial preventative effects.

MIRANDA REED

I started my career studying toxicology and was primarily interested in how early-life exposure to toxicants, such as methylmercury, accelerates the aging process later in life. To examine this, rats were gestationally exposed to methylmercury and assessed using a battery of motor and cognitive tasks throughout life. As I continued my training, I realized the aging portion of the project, more so than the toxicology or early-life portions, interested me most. After receiving my Ph.D., I sought a postdoctoral position that focused on aging or age-related diseases and ultimately was hired at the University of Minnesota as a postdoctoral fellow in Karen Ashe's laboratory. There, my work focused on the molecular and behavioral mechanisms underlying Alzheimer's disease using transgenic mouse models. Since being appointed to a faculty position at West Virginia University, my work has focused on the mechanism by which various factors, such as stroke and diabetes, increase the risk for Alzheimer's disease. Through this work, I became acquainted, and began collaborating with, Dr. James Simpkins in traumatic brain injury, stroke, Alzheimer's disease, and the neuroprotective effects of estrogen. Of utmost importance, as I see it, is to identify and understand the parameters under which estrogen is beneficial.

COLIN J. SALDANHA

My interest in the hormonal modulation of brain and behavior began as an undergraduate student at Gustavus Adolphus College, and continued as a graduate student at Columbia University. It was during the 1980s and 1990s when we learned that estrogen was not just a "female hormone," but was critically important in organizing and activating the brains of both sexes. Perhaps more importantly, the range of physiological endpoints affected by estrogen spanned much more than reproductive behavior. This range included a role for estrogens in protecting the brain from insult caused by trauma and or aging. While studying the expression of the enzyme responsible for estrogen synthesis as a postdoctoral trainee at UCLA, we discovered that this enzyme was induced in reactive astrocytes around the site of mechanical damage. We have spent the last two decades trying to characterize and understand how and why this occurs in songbirds, a group of vertebrates with documented neural sensitivity to estrogens and dramatic neuroplasticity. We have learned quite a bit about this phenomenon, including some striking similarities and differences in the induction and consequences of astrocytic aromatization in birds and mammals. Importantly, there are more similarities than differences.

JAMES SIMPKINS

My interest in estrogen effects on the brain began with my Ph.D. training at the Michigan State University in the laboratory of the pioneering neuroendocrinologist, Dr. Joseph Meites. There, I received superb training in the mechanisms by which the brain regulates reproductive function through its control of the pituitary gland. Following my first appointment to a faculty position at the University of Florida, I became interested in the effects of ovarian hormones on brain processes, including mood, cognition, hot flashes, energy metabolism and food intake. A Master of Science student in my laboratory, Jean Bishop, made the serendipitous discovery that when she deprived neurons of nutrients to kill them, estrogens potently protected cell (Bishop & Simpkins, 1994). This nearly forgotten discovery launched the field of estrogen neuroprotection and resulted in hundreds of subsequent confirmatory publications, and has defined the rest of my career focus. Our current effort, and that of the rest of the field, to determine the mechanism(s) of estrogen neuroprotection in traumatic brain injury, stroke, Alzheimer's disease and Parkinson's disease, are an extension of this cornerstone discovery made by a bright young graduate student.

FARIDA SOHRABJI

My interest in estrogen effects on the brain began with my graduate training at the University of Rochester. While estrogen has well-known effects on reproduction, I was interested in the less well-known effects of this hormone on brain functions such as learning, memory and emotion. At the University of Rochester (NY), I trained with Drs. Kathy and Ernest Nordeen and studied song regions in the zebra finch brain, which is a hormone-sensitive circuit. I received postdoctoral training with Dr. C. Dominique Toran-Allerand at Columbia University (NY).

Dr. Toran-Allerand was a pioneer in the study of hormone effects on the brain and was the first to show that estrogen directly influenced the developing brain. During the same time, the first studies were published showing a beneficial effect of estrogen in neurodegenerative disease. I was very intrigued by these studies and pursued this line of research when I obtained my first faculty position. Like many other labs, we found that estrogen was neuroprotective and anti-inflammatory but a turning point in my research came with a talented graduate student (Melinda Jezierski). She observed that, while estrogen treatment in young females was neuroprotective and anti-inflammatory, estrogen treatment in middle-aged female rats had adverse outcomes. This observation gave rise to a series of studies designed to understand why estrogen treatment was beneficial to young animals but damaging to older animals, and continues to influence our current research on stroke therapies.

Introduction

> Life is a struggle, not against sin, not against the Money Power, not against malicious animal magnetism, but against hydrogen ions. —*H.L. Mencken*

> Never underestimate the power of estrogen. —*Unknown*

WHY SHOULD I CARE?

From car accidents to pee-wee football, injuries to the brain are a hot topic—so much so that many states and the federal government are considering, or implementing, laws in direct response to suspected brain injury. The U.S. Centers for Disease Control and Prevention (CDC) states that sports and recreational activities alone account for over 200,000 visits to emergency rooms for concussion and other brain injuries annually. Additionally, the U.S. Department of Defense reports that traumatic brain injury (TBI) is one of the invisible wounds of war and one of the signature injuries of troops wounded in Afghanistan and Iraq. Each year in the United States, traumatic brain injury affects about 270,000 armed service members. Together, these types of injuries affect almost 2 million people each year, and many are fatal. Thus, identifying potential factors that mediate the extent or repair of damage is of great importance.

Symptoms of TBI can be mild to severe and, while some symptoms are evident immediately following injury, others do not surface until several days to many weeks later. TBI has become a great concern in terms of public health and disease, and treatment options. Because the effects of TBI can be both acute (inflammation and cell death) and prolonged (decline in cognitive function and ability), understanding the mechanisms that mediate neuroprotection and repair are imperative.

Prevention of injuries is key; thus, many aspects of research are focused on decreasing the chances of sustaining injuries. This is done through increased safety standards in vehicles and new and improved helmets in sporting and recreational activities. Researchers are also trying to develop better models for studying the varying types of TBI. However, research also focuses on targeting mechanisms of neuroprotection and repair after an injury has occurred. These include, but are not limited to, targeting blood–brain barrier repair and dynamics, decreasing reactive oxygen species and other metabolites produced following injury, as well as the complex neuroinflammatory response that occurs. Towards this goal, estrogens have emerged as a potential theurapeutic agent in decreasing the damaging effects of TBI.

Estrogens are traditionally thought to be involved in the classical model of brain sexual differentiation and, with the gonads, act to initiate male or female brain development. However, research increasingly suggests that estrogens have profound effects on neuroprotection and repair following damage to the brain.

RATIONALE AND SCOPE OF THE BOOK

The goal of this book is to demystify, deconstruct, and humanize the field of estrogen-mediated neuroprotection following TBI, and make the subject approachable to both undergraduates and graduate students. This book brings together leading researchers and practitioners to explain the basis for their work, methods, and results. Chapters explore the history of traumatic brain injury, the structure and function of estrogens and the estrogen receptor, as well as a comprehensive review of what is known about the role of estrogens following damage to the brain, ranging from penetrating injuries to stroke. Topics include induction of estrogen response following injury, consequences of estrogen action, and mechanisms underlying estrogen-mediated neuroprotection. This book will be of importance to teachers, researchers, and clinicians interested in the role that estrogens play following traumatic brain injury. This book is also unique in that the same researchers and practitioners will provide a look into how they got into research and their scientific field.

CHAPTER

1

Historical Antecedents

John P. Hayden

Biology, Vassar College, Poughkeepsie, New York, USA

INTRODUCTION

On September 13, 1848, 25-year-old Phineas Gage arrived at the doorstep of Vermont physician Dr. John Harlow after an accidental explosion at a railroad construction site had catapulted a 1-meter long, 6-kg iron tamping rod into Gage's cheek, through his brain, and out the top of his skull (Kotowicz, 2007; Figure 1.1). Incredibly, the young man had not lost consciousness, and was lucid enough to recount the tale to a likely dumbfounded Harlow. The improbable story of Phineas Gage, a favorite among neuroscientists, has become infamous. From the shocking lack of immediate symptoms to the accounts of drastic personality change and recovery, all aspects of his tale grip our imaginations and probe us to ask the question:

How did Phineas Gage survive such a horrific brain injury?

It has now been more than 150 years since Gage's accident, and we are still learning about the consequences of trauma to the brain. In this chapter, we will explain the most recent definitions of traumatic brain injury (TBI), and explore how our understanding of TBI and its effective treatment have changed from ancient times to the 21st century. Finally, we will examine how both basic scientists and clinicians have revolutionized how we treat the brain. In particular, we will focus on the role of steroid hormones, specifically estrogens, as effective tools in the current management of TBI.

THE "INS AND OUTS" OF TBI: AN OVERVIEW

Not all injuries that inflict trauma to the brain are the same. For this reason, our understanding of modern TBI terminology will begin with

K.A. Duncan (Ed): *Estrogen Effects on Traumatic Brain Injury.*
DOI: http://dx.doi.org/10.1016/B978-0-12-801479-0.00001-2

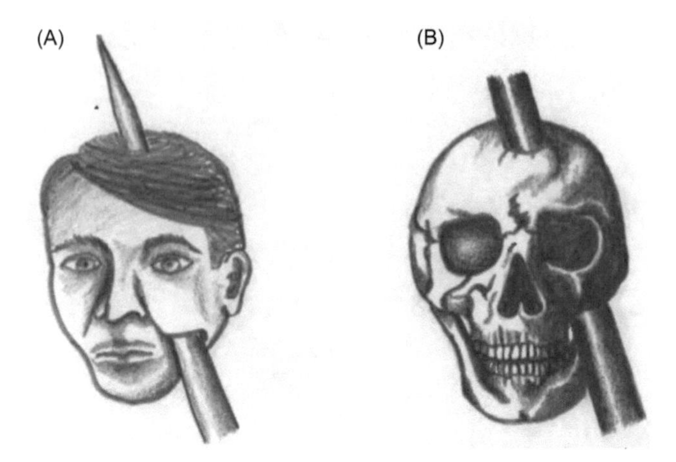

FIGURE 1.1 Images of Phineas Gage based upon three figures from Harlow's 1868 paper. A) Depicts the angle of the rod through Phineas Gage's head. B) Image depicting the damage and injury tract to Phineas Gage's skull and frontal lobe following the accident.

an explanation of three broad categories, which we will use to divide the injuries by mechanisms of assault (Table 1.1). The most common category of injury is **blunt force trauma** or "closed-head injury." Blunt force trauma often presents as a concussion, intracranial hematoma, cerebral contusion, or a diffuse axonal injury. Despite serious damage to the brain, skull and dura mater remain physically intact. Examples of blunt force injuries include vehicular accidents, falls, acts of violence, and sports injuries. Blunt force trauma can range from single/multiple mild incidents (as observed in most sports-related injuries) to more severe blast-related incidents (as observed in recent veterans of the various wars in the Middle East). These types of injuries can result in major neurological and/or cognitive deficits or even fatality (Reilly and Bullock, 2005; Taber et al., 2006).

The second category of injury, **penetrating injury**, occurs when a foreign object pierces the nervous tissue, specifically the brain, causing localized damage along the path of entry (Reilly and Bullock, 2005). The most common presentations of a penetrating injury are related to gun violence and stabbings. In the case of penetrating injuries, both the skull and the dura mater are damaged. Penetrating injuries can result in shear-like injury to the neurons, epidural hematomas, subdural hematomas, or parenchymal contusions. Though less common than blunt force trauma, penetrating injuries carry a far worse prognosis (Kazim et al., 2011). While these first two categories are useful tools in qualifying injuries beyond "head trauma," they are not mutually exclusive. It is

TABLE 1.1 Categories of Traumatic Brain Injury

	Definition	Possible symptoms	Clinical examples
Blunt Force Trauma	Non-penetrating impact to head	Concussion, intracranial hematoma, cerebral contusion, diffuse axonal injury, neurological and/or cognitive deficits	Vehicular accidents, falls, acts of violence, and sports injuries
Penetrating Injury	Damage that results from a foreign object penetrating the nervous tissue	Shear-like injury to neurons, epidural hematomas, subdural hematomas, parenchymal contusions	Gun shot wounds, stabbing
Hypoxic-ischemic brain injury	Damage caused by a reduction in oxygen and/or diminished blood supply	Seizures, disturbances of sensorimotor function, neuronal death, neurological and/or cognitive deficits	Cardiac arrest, respiratory arrest, near-drowning or hanging, carbon monoxide exposure, and perinatal asphyxia

entirely possible for a single accident to result in both blunt force and penetrating injuries. After the iron rod shot through Gage's skull, he fell to the ground, and it is very likely that, in addition to the damage caused by the rod, a secondary impact occurred. In cases like this, the brain would exhibit both localized and global injuries from the penetrating and blunt force injuries, respectively.

The final category of TBI that will be explored in this book is not caused by forces against the brain at all but instead is the result of decreased glucose and oxygen. **Hypoxic-ischemic brain injury** is any injury to the brain induced by hypoxia (reduction in oxygen) and ischemia (diminished blood supply; Busl and Greer, 2010). The typical causes of hypoxic-ischemic brain injury include cardiac arrest, respiratory arrest, near-drowning or hanging, carbon monoxide exposure, and perinatal asphyxia, and resulting symptoms include seizures, disturbances of sensorimotor function, neuronal death, and neurological and/or cognitive deficits (Busl and Greer, 2010).

Beyond these rudimentary classifications, two distinct stages of injury occur within all three mechanical subtypes of TBI (Table 1.2). The first stage of damage is referred to as the **primary injury** because it occurs at the time of assault and is the direct result of trauma to the brain (Reilly and Bullock, 2005). This initial trauma can include

TABLE 1.2 Definition and Components of Two Stages of Traumatic Brain Injury

	Definition	Components
Primary Injury	Trauma to the brain that is localized to the site of injury, and occurs at the time of assault	Tearing, shearing and rupturing of blood vessels, direct damage to brain tissue, neurons and blood–brain barrier
Secondary Injury	Slowly developing damage that results from biochemical and physical responses to the initial trauma	Ischemia, hypoxia, cerebral edema, oxidative damage, glucose starvation, acidosis

lacerations, contusions, hematomas, sheared neurons, and fractures of the skull (Marion, 1999). Because the primary injury is only defined by the initial physical trauma, any symptoms of the primary injury may be immediately apparent following assault. Phineas Gage was able to get up and walk away from his accident with relatively minor impairment because his primary injuries miraculously did not cause enough physical damage to impinge on his basic bodily functions. Nerves and tissue had certainly been damaged, but likely manifested their trauma in less severe symptoms such as nausea, dizziness and transient loss of consciousness.

Gage's incredible resilience was only temporary, however, and following the accident his mental and physical health declined. The apparent discontinuity between his acute and chronic symptoms was likely caused by the onset of the next stage of TBI, the **secondary injury**. The secondary injury develops in the aftermath of assault, and includes various detrimental biochemical and physical changes within the brain (Coetzer, 2006). In contrast to the primary injury, the damage caused by secondary injury develops slowly, worsening in the aftermath of the assault. Deviations from normal physiological levels of acetylcholine and free radicals in the brain can combine with other biochemical changes to restrict cerebral blood flow following injury (Coetzer, 2006). Further restriction may also result from pro-inflammatory pathways that cause an increase in neurodegenerative properties, such as edema. As the brain expands beyond its normal volume, it may also press itself against the rigid boundaries of the skull, resulting in both increased intracranial pressure (ICP), and additional restriction of blood flow (Coetzer, 2006). The ischemia caused by this reduction causes a damaging deficit in the amount of oxygen and glucose that are available to the brain. These various effects compound one another, and result in increased damage to the brain. For Gage, it is very possible that his symptoms were exacerbated in the days following his injury by increasing edema, ischemia, and axonal injury. Over the course of several days, an injury that once allowed Gage to immediately rise to his feet and speak with the doctor likely became much worse, and eventually Gage's own body would have left itself with a more severe injury than was initially caused by the rod.

Gage's reported decline following his injury highlights the critical component of the secondary injury that makes it so important to contemporary TBI research. The secondary injury represents the "access point" that doctors have to limit the amount of damage that occurs following injury. While the primary injury cannot be undone, scientists can attempt to hinder the progress of the secondary injury before it has the opportunity to inflict further damage.

A HISTORY OF TBI FROM ANTIQUITY THROUGH THE 19TH CENTURY

Today, it is not uncommon to turn on the TV or scroll down a news site and see a story about TBI. From the NFL and the military, to playgrounds in schools across the country, discussions of TBI risk and prevention are increasingly a part of our everyday vernacular. While this trend has rightfully increased general awareness of such an important issue, it is by no means the beginning of TBI's story in the human narrative. The significance of the topic therefore extends far beyond our modern day news, all the way back into the earliest annals of human history.

Some of the first accounts of traumatic brain injury are thought to predate our written history, appearing in ancient myths that were passed down orally (Courville, 1967). By about 1550 BC, TBI was considered important enough to appear on one of the earliest known medical documents, the "Edwin Smith Papyrus" (Nunn, 2002). The scroll, which is the first place the word "brain" appears in medical literature (Sanchez and Burridge, 2007), contains instructions for ancient Egyptian doctors on how to examine, diagnose, and treat traumatic head injuries. For an open fracture of the skull (case #20 on the papyrus) the document states that the patient should be "...sat up, his head softened with grease, and milk put into his ears" (Nunn, 2002). While some aspects of the treatments may seem comic today, others remain quite valid, such as using blood in the nose and ears as indicators for a fractured base of the skull (Nunn, 2002).

Within ancient Mesopotamia, the understanding of what symptoms accompany TBI was even more refined. For these war-prone civilizations, head injuries may have occurred quite frequently, and the resulting loss of hearing, seizures and paralysis they described may have been all too common (Scurlock and Andersen, 2005). For those who survived long enough to receive treatment, the procedure of choice would likely have been trepanation (Castiglioni, 1958), a surgical intervention in which a hole is drilled or scraped into the skull, exposing the dura mater. The usage of trepanation was widespread for a variety of intracranial ailments, and ancient trepanned skulls have been found throughout the world (Levin, 1982). It remained a popular, though likely ineffective, surgical treatment into medieval times. In particular, 12th century surgeon Roger of Salerno incorporated modifications of the procedure based on additional diagnostic criteria (Castiglioni, 1958). Flash forward to the 15th and 16th centuries, and Renaissance surgeons continued to refine these practices, as they began to deal with more modern causes of TBI such as gunshot wounds (Levin, 1982; Gurdjian, 1973). TBI research then slowed to a sluggish pace, which continued into the 19th century, as the ideas of researchers likely outpaced advances in technology that would have allowed for investigation. Despite these challenges, however, scientists in the 19th century did manage to lay the foundations for the identification of mechanisms underlying secondary injury, even describing the degeneration of nervous tissue and cells surrounding injury (Levin, 1982).

FROM THE BENCH TO THE CLINIC: TBI IN THE 20TH AND 21ST CENTURY

Despite TBI's long existence within the human narrative, its treatment has been the product of a relatively short window of basic science research (High et al., 2005). Most penetrating brain injuries suffered prior to the 1900s were fatal (Gurdjian, 1973). Today, overall death rates in patients with severe traumatic brain injury have been reported to be as low as 13% overall (Gerber et al., 2013). After nearly 100 years of treatment stagnation throughout the 19th century, the invention of the intracranial pressure monitor in the 1950s improved doctors' ability to treat TBI (Lundberg et al., 1959; Lundberg et al., 1965). Even today, ICP monitoring remains a fundamental component of TBI patient care (Gerber et al., 2013). Additional technological advancements furthered clinical research with the invention and refinement of CT scanning in the 1960s and 1970s. The CT scan allowed for revolutionary visualization of the injuries in living patients, and soon became standard practice for the evaluation of TBI (Marion, 1999). First cautious use of hormones as a treatment for TBI began around this time as well. Discoveries by Galicich and French (1961) suggested that the administration of steroids could reduce the elevation in intracranial pressure associated with brain tumors. Although these early findings caused the rapid adoption of steroid treatment for TBI patients, subsequent studies indicated that the steroids, while effective with tumors, had no effect on the ICP of TBI patients. These studies emphasized the need for researchers to clarify the unique biological mechanisms of TBI, as well as the roles that hormones played within them (Marion, 1999).

In the 1980s, scientists began to parse out these details. As stated earlier, researchers had postulated as to the mechanisms of secondary injury as early as 1835, some even showing degeneration of nerves surrounding the injury (Levin, 1982). But these findings were left relatively undeveloped until technological advancements throughout the 20th century made further investigation possible. In addition to the new and improved imaging techniques, novel models of replicating TBI in a lab setting were pioneered during this time (Marion, 1999). Together, these advances not only uncovered changes in the metabolic demands of the brain following TBI, but also clarified the roles of inflammatory and oxidative damage in secondary injury (Marion, 1999). These discoveries along with other revelations about the molecular and physiological processes of secondary injury directly impacted clinical practices. The use of hyperventilation as a treatment was shown to be ineffective following research on the metabolic changes following TBI, and treatments that improve cerebral blood flow became increasingly popular to avoid ischemia (Marion, 1999).

While clinicians tried to keep pace with the advancements in laboratory science, the absence of a standardized system to classify aspects of TBI severity and recovery had become apparent. Clinicians began to use the Glasgow Coma and Outcome Scales, which together numerically classify the severity of TBI and quality of subsequent outcome (Marion, 1999). These scales revolutionized not only patient care, but also researchers' abilities to assess the prevalence, incidence and demographics of TBI.

One of the revolutionary discoveries that followed the implementation of the Glasgow scales was the elucidation that "sex matters" following TBI. Early studies examining the prevalence of stroke demonstrated that premenopausal women experience fewer strokes than men of comparable age. However, stroke rates and severity increase among postmenopausal women compared with age-matched men (Giroud et al., 1991; Sacco et al., 1998). An explosion in sex-difference driven research followed. In the 1990s and early 2000s researchers demonstrated a vast array of sex-dependent differences following TBI. By 1998, researchers had identified marked differences in the recoveries of females and male patients following TBI (Herson et al., 2009). It was shown in mice and rats that females experienced less tissue damage from equivalent injuries than their male counterparts (Alkayed et al., 1998; Alkayed et al., 2000; Carswell et al., 1999; Hall et al., 1991) and that their functional outcome was improved (Li et al., 2004). Progesterone, a sex steroid, was hypothesized to play a pivotal role in these differences (Groswasser et al., 1998). At the same time, groundbreaking research on the presence and production of neurosteroids was flourishing (Herson et al., 2009). Furthermore, the discovery of adult neurogenesis and the role of estrogens in mediating these processes became known (Doupe, 1994; Goldman, 1998; Nottebohm, 1985; Sohrabji et al., 1994). Together, these revelations caused a refocusing within the research community on the possibility of sex steroids playing a critical role in TBI recovery and new steroid based TBI therapies began (Stein, 2001).

CONCLUSIONS AND FUTURE THOUGHTS

Over 150 years after Phineas Gage's unfortunate accident we are still trying to discover how to repair the brain. Furthermore, we know that injuries once considered minor can have very profound effects on brain tissue and behavior. In order to develop these treatment options we must 1) maintain the interactions between basic scientists and clinicians in order to develop translational treatment options, 2) develop animal models that most effectively mirror TBI in humans, 3) determine what, if any, effect sex and gender have on TBI outcome, and 4) determine the effect of age on TBI outcome.

Acknowledgements

The authors would like to thank John M. Duncan for illustrations used within this chapter.

References

Alkayed, N.J., Harukuni, I., Kimes, A.S., London, E.D., Traystman, R.J., Hurn, P.D., 1998. Gender-linked brain injury in experimental stroke. Stroke. 29 (1), 159–166.

Alkayed, N.J., Murphy, S.J., Traystman, R.J., Hurn, P.D., 2000. Neuroprotective effects of female gonadal steroids in reproductively senescent female rats. Stroke. 31 (1), 161–168.

Busl, K.M., Greer, D.M., 2010. Hypoxic-ischemic brain injury: pathophysiology, neuropathology and mechanisms. NeuroRehabilitation. 26 (1), 5–13.

Carswell, H.V., Anderson, N.H., Clark, J.S., Graham, D., Jeffs, B., Dominiczak, A.F., et al., 1999. Genetic and gender influences on sensitivity to focal cerebral ischemia in the stroke-prone spontaneously hypertensive rat. Hypertension. 33 (2), 681–685.

Castiglioni, A., 1958. A History of Medicine. Kessinger Publishing, Whitefish, Montana.

Coetzer, R., 2006. Traumatic Brain Injury Rehabilitation: A Psychotherapeutic Approach to Loss and Grief. Nova Science Publishers, Inc., New York.

Courville, C.B., 1967. Injuries of the Skull and Brain: As Described in Myths, Legends and Folktales of the Various Peoples of the World. Vantage Press, New York.
Doupe, A.J., 1994. Songbirds and adult neurogenesis: a new role for hormones. Proc. Natl. Acad. Sci. USA 91 (17), 7836.
Galicich, J.K., French, L.A., 1961. Use of dexamethasone in the treatment of cerebral edema resulting from brain tumors and brain surgery. Am. Pract. 12, 212–223.
Gerber, L.M., Chiu, Y.L., Carney, N., Hartl, R., Ghajar, J., 2013. Marked reduction in mortality in patients with severe traumatic brain injury. J. Neurosurg. 119 (6), 1583–1590.
Giroud, M., Milan, C., Beuriat, P., Gras, P., Essayagh, E., Arveux, P., et al., 1991. Incidence and survival rates during a two-year period of intracerebral and subarachnoid haemorrhages, cortical infarcts, lacunes and transient ischaemic attacks. The Stroke Registry of Dijon: 1985–1989. Int. J. Epidemiol. 20 (4), 892–899.
Goldman, S.A., 1998. Adult neurogenesis: from canaries to the clinic. J. Neurobiol. 36 (2), 267–286.
Groswasser, Z., Cohen, M., Keren, O., 1998. Female TBI patients recover better than males. Brain Injury. 12 (9), 805–808.
Gurdjian, E.D., 1973. Head Injuries from Antiquity to the Present with Special Reference to Penetrating Head Wounds. Charles C. Thomas, Springfield, IL.
Hall, E.D., Pazara, K.E., Braughler, J.M., 1991. Effects of tirilazad mesylate on postischemic brain lipid peroxidation and recovery of extracellular calcium in gerbils. Stroke. 22 (3), 361–366.
Herson, P.S., Koerner, I.P., Hurn, P.D., 2009. Sex, sex steroids, and brain injury. Semin. Reprod. Med. 27 (3), 229–239. NIH Public Access.
High, W., Sander, A.M., Struchen, M.A., Hart, K.A., 2005. Rehabilitation for Traumatic Brain Injury, Baylor College of Medicine. Oxford University Press.
Kazim, S.F., Shamim, M.S., Tahir, M.Z., Enam, S.A., Waheed, S., 2011. Management of penetrating brain injury. J. Emerg. Trauma Shock. 4 (3), 395.
Kotowicz, Z., 2007. The strange case of Phineas Gage. Hist. Human. Sci. 20 (1), 115–131.
Levin, H.S., 1982. Neurobehavioral Consequences of Closed Head Injury. Oxford University Press.
Li, X., Blizzard, K.K., Zeng, Z., DeVries, A.C., Hurn, P.D., McCullough, L.D., 2004. Chronic behavioral testing after focal ischemia in the mouse: functional recovery and the effects of gender. Exp. Neurol. 187 (1), 94–104.
Lundberg, N., Kjallquist, A., Bien, C., 1959. Reduction of increased intracranial pressure by hyperventilation. Acta Psychiatr. Scand. 34, 4–64.
Lundberg, N., Troupp, H., Lorin, H., 1965. Continuous recording of the ventricular fluid pressure in patients with severe acute traumatic brain damage. A preliminary report. J. Neurosurgery. 22, 581–590.
Marion, D., 1999. Traumatic Brain Injury. Thieme, New York.
Nottebohm, F., 1985. Neuronal replacement in adulthood. Ann. N. Y. Acad. Sci. 457 (1), 143–161.
Nunn, J.F., 2002. Ancient Egyptian Medicine. University of Oklahoma Press, Norman, OK.
Reilly, P.L., Bullock, R. (Eds.), 2005. Head Injury: Pathophysiology and Management. Hodder Arnold, London.
Sacco, R.L., Boden-Albala, B., Gan, R., Chen, X., Kargman, D.E., Shea, S., et al., 1998. Stroke incidence among white, black, and Hispanic residents of an urban community: the Northern Manhattan Stroke Study. Am. J. Epidemiol. 147 (3), 259–268.
Sanchez, G.M., Burridge, A.L., 2007. Decision making in head injury management in the Edwin Smith Papyrus. Neurosurg. Focus. 23 (1), E5.
Scurlock, J.A., Andersen, B.R., 2005. Diagnoses in Assyrian and Babylonian Medicine: Ancient Sources, Translations, and Modern Medical Analyses. University of Illinois Press, Urbana, 307.

Sohrabji, F., Miranda, R.C., Toran-Allerand, C.D., 1994. Estrogen differentially regulates estrogen and nerve growth factor receptor mRNAs in adult sensory neurons. Obstet. Gynecol. Surv. 49 (7), 495–497.

Stein, D.G., 2001. Brain damage, sex hormones and recovery: a new role for progesterone and estrogen? Trends. Neurosci. 24 (7), 386–391.

Taber, K., Warden, D., Hurley, R., 2006. Blast-related traumatic brain injury: what is known? J. Neuropsychiatry. Clin. Neurosci. 18 (2), 141–145.

CHAPTER

2

Estrogen Actions in the Brain

Laura L. Carruth and Mahin Shahbazi

Neuroscience Institute, Georgia State University, Atlanta, Georgia, USA

INTRODUCTION

Across an animal's lifespan the brain is continuously exposed to chemical factors, such as hormones, that travel via the bloodstream, to influence various biological processes. One such process influenced by bloodborne hormones is behavior. The hormonal regulation of brain and behavior in vertebrates has fascinated generations of scientists, in particular the knowledge that, while the functions of different steroids have changed multiple times over evolutionary time, their chemical structures are still highly conserved and nearly identical among different vertebrate animals.

The bidirectional interaction between hormones and behavior, as well as the understanding that hormones do not directly alter behavior, but only change the probability that a behavior can occur within an appropriate social and behavioral context, intrigued us both and formed the foundation of our work as behavioral neuroendocrinologists. While there is abundant information from the last century or so on the functions of steroids in regulating reproductive behavior, it was the exciting research on the role of estradiol (E2) in sexual differentiation of the mammalian brain that captivated both of our interests. This work was fundamentally grounded in findings that demonstrated that testosterone (T) is converted into E2 in the brain via the enzyme *P*-450 aromatase, and that it is E2 that promotes masculinization of the brain. This research was equally captivating because it suggested that the brain can function as an endocrine gland, and that local hormone synthesis can influence brain development and behavior, and that these actions were not just under the control of hormones secreted from the gonads. Within the last decade exciting new advances in the field have allowed researchers to measure

K.A. Duncan (Ed): Estrogen Effects on Traumatic Brain Injury.
DOI: http://dx.doi.org/10.1016/B978-0-12-801479-0.00002-4

neuroendocrine changes at the genetic, epigenetic, transcriptomic, and proteomic levels, and the integration of this data will form the basis of neuroendocrine research in the years ahead.

Estrogens are steroid hormones that have been well studied for the major role they play in female development and reproductive behavior. There is almost a century of research investigating just how estrogens regulate female reproductive physiology and behavior (see Boling and Blandau, 1939), and in more recent decades research has examined how estrogens act in the brain to regulate behaviors beyond female reproduction. Early studies explored how female reproductive behavior could be disrupted after the removal of the ovaries (Beach, 1948) and how prenatal steroid hormone treatment either rescued or resulted in abnormal sexual development (Phoenix et al., 1959). The actions of estradiol have been extensively explored within the context of the classical action of steroids involving the secretion from a steroidogenic endocrine gland, traveling via the bloodstream, and then acting on a distant tissue expressing the appropriate receptors, in particular in the brain. Estrogens acting in this "classical" manner are considered to be neuroactive steroids. However, it is now well known that estradiol can also be synthesized *de novo* by the developing brain, and thus is classified as a neurosteroid. Estrogens produced as neurosteroids may be involved in rapid estrogen signaling, acting within minutes instead of hours to days, which is seen when they act through classical mechanisms. This chapter focuses on the actions of estrogens in the mammalian brain. We will touch on the biochemistry of estrogens; the mechanisms of estradiol action in the brain including rapid signaling, sexual differentiation of the brain, and examples of brain regions or neurotransmitter systems that are differentially affected by estrogens; and the varying actions of estrogens in aging and neuroprotection.

STEROID HORMONES

Steroid hormones are a major class of hormones characterized as being fat-soluble organic compounds that can easily pass through cell membranes. Due to being fat-soluble, steroids are secreted from cells as soon as they are synthesized from the smooth endoplasmic reticulum and inner membrane of mitochondria and are never stored in intracellular vesicles. Circulating steroids can also easily cross the blood–brain barrier (BBB) and act in the brain (Lovejoy, 2005). Because steroids are not water-soluble they are generally bound to carrier proteins, such as sex hormone binding globulins (Caldwell and Jirikowski, 2014), in blood to facilitate transport to target tissues expressing the appropriate receptors. The precursor for all steroid hormones is cholesterol, made from acetate in the liver. The chemical structure of steroids includes three cyclohexane rings plus one conjugated cyclopentane ring. Steroids are nonpolar, resulting from the large number of carbon-hydrogen bonds.

Estrogens

Estrogens are classified as C_{18} steroids, a versatile group of hormones that includes estrone, estriol, and the most biologically prevalent and potent of the estrogens, 17β-estradiol (or E2). The precursors for all estrogens are androgens (such as testosterone, T, and androstenedione). The steroidogenesis of androgens to estrogens requires the enzyme *P*-450 aromatase (also known as estrogen synthetase or estrogen synthase) to remove the carbon at position 19, via a process called aromatization, resulting in an aromatic estrogen compound with a phenolic A-ring (see Figure 2.1). Aromatase is a member of the cytochrome P450 superfamily of monoxygenases and is expressed in the gonads, brain, adrenal cortex, adipose tissue, bone, skin, and blood vessels (Harada et al., 1993). While T is aromatized to E2, androstenedione is primarily converted to estrone. Some androgens like dihydrotestosterone (DHT) cannot be aromatized into an estrogen. Ovarian granulosa cells produce large amounts of androgens, which are immediately converted to estrogens which can be released into the bloodstream. The actions of estrogens result in several important phenotypic changes that include, but are not limited to, oogenesis or the production of eggs, functions related to feeding and reproduction, development of female secondary sexual morphological characteristics, and various aspects of female reproductive behavior, as well as cognition, memory, and aspects of neuroprotection. At the basic level of functioning, the physiological impacts of estrogens involve two processes: hormone binding through nuclear or membrane associated receptors that upregulate specific biochemical pathways via

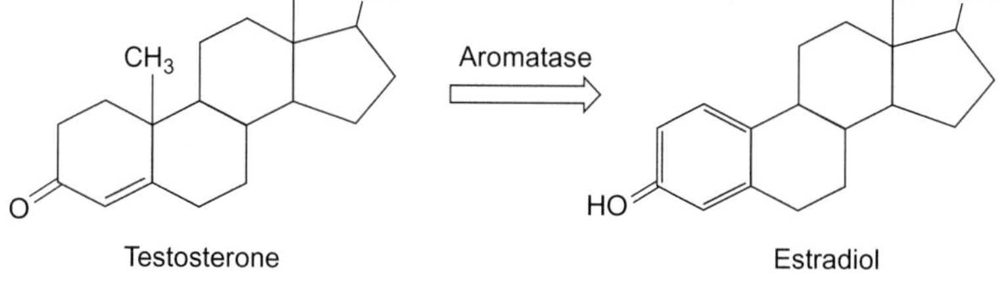

CH_3 OH CH_3 O Testosterone Aromatase CH_3 OH HO Estradiol

FIGURE 2.1 Testosterone is converted into estradiol via the enzyme aromatase.

intracellular machinery to affect protein translation, and the endogenous regulation of ovarian estradiol synthesis via negative and positive feedback loops.

Estrogen Receptors

One mechanism of action for estrogens is via binding to the appropriate intracellular or extranuclear receptor. Long-term steroid effects are most often mediated via classical genomic processes, and regulated via negative feedback loops, while short-term, rapid effects can be mediated via nongenomic mechanisms (see Figure 2.2). Estrogen receptors (ERs) are members of the superfamily class of nuclear receptors located in either the cell cytoplasm or nucleus and which function as transcription factors (Lovejoy, 2005). There are currently two recognized ERs that function as nuclear receptors, ERα (NR3A1) and ERβ (NR3A2) (Beato and Klug, 2000). In their unbound state, inactive ERs are present in the cytoplasm. After ligand binding, ERs form ERα (αα) or ERβ (ββ)

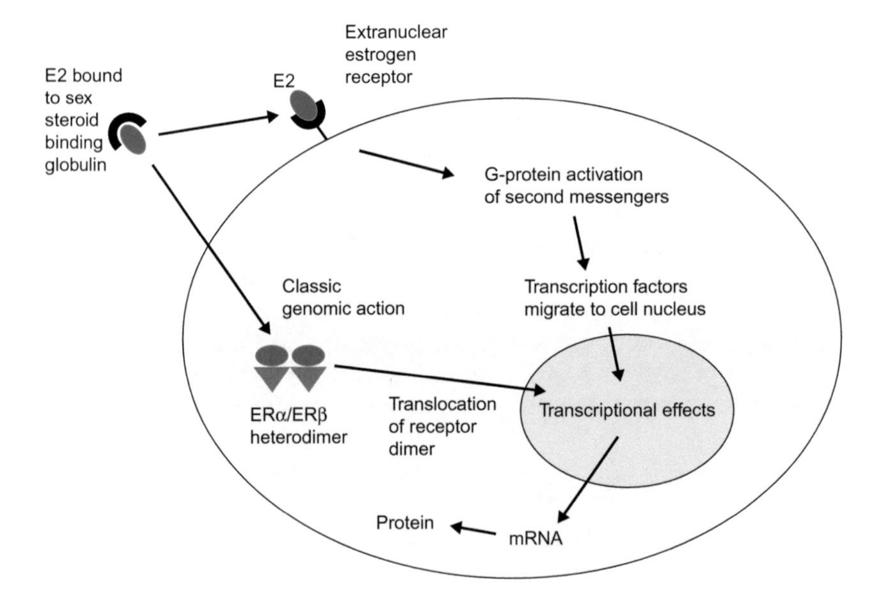

FIGURE 2.2 Presumed mechanisms of estrogen action via intracellular and extranuclear ERs. Direct genomic responses involve ER receptor dimerization, translocation into the nucleus, and binding with ERE. Coactivator proteins and RNA polymerases are recruited by the receptor-DNA complex to facilitate mRNA transcription, and then the subsequent translation into protein. Nongenomic E2 action involves binding to a cell surface receptor, activation of G-proteins, followed by rapid activation of kinase signaling pathways.

homodimers or ERα-ERβ (αβ) heterodimers and translocate into the nucleus. ERs are characterized by having a two zinc finger DNA-binding domain at the amino terminus, a central region containing a nuclear localization signal, and a ligand-binding region at the carboxy terminus (Lovejoy, 2005) with ERα and ERβ only sharing 56% homology in the ligand-binding domain. The DNA-binding domain targets the ER to an estrogen response element (ERE), and it is the ERE that binds an ER homo- or heterodimer. Moreover, many different coactivator or corepressor proteins are recruited as part of the transcriptional regulatory complex in E2 induced gene expression (Tetel, 2009).

The brain is a target for circulating steroids due to the large number of ER present in the brain; however, the distributions of ERα and ERβ in the body differ quite distinctly. There is high ERα expression in the pituitary, kidney, epididymis, and adrenal while ERβ is prominently expressed in prostate, lung and bladder. They share overlapping expression in brain, ovary, uterus, and testis (Kuiper et al., 1998). A thorough analysis of ERα and ERβ protein in the postnatal rat brain by Pérez and colleagues (2003) demonstrates a wider distribution for ERα than ERβ. Both receptor subtypes were detected in different areas of the cerebral cortex, basal forebrain, amygdala, thalamus, hypothalamus, mesencephalon, pons, cerebellum, and medulla oblongata, while hippocampal staining was exclusively ERα. Looking in detail at the rat hypothalamus, abundant ERα mRNA expression was observed in the arcuate and ventromedial nuclei, while ERβ was readily expressed in the anterior periventricular, paraventricular, supraoptic, medial tuberal, and medial mammillary nuclei. Both the medial preoptic area of the hypothalamus and the bed nucleus of the stria terminalis express both ERα and ERβ (Shughrue et al., 1996).

Beyond the genomic actions attributed to estrogens, there is much evidence that suggests a rapid signaling mechanism for E2 mediated via extranuclear ERs. There are a number of putative extranuclear membrane ERs, including ER-X and G-protein coupled estrogen receptor-30 (GPR30), as well as membrane-bound splice variants of ERα and ERβ. Via extranuclear receptors, E2 can exert effects within minutes (Tang et al., 2014). GPR30 is a 7-transmembrane domain receptor that has been localized in the plasma membrane, Golgi, endoplasmic reticulum, as well as dendritic spines in the brain (Hazell et al., 2009; Tang et al., 2014). This receptor was first recognized in breast cancer cell lines but has subsequently been shown to be expressed in the plasma membranes of neurons in the paraventricular nucleus and supraoptic nucleus of the hypothalamus, in the anterior and posterior pituitary, hippocampus (Brailoiu et al., 2007; Hazell et al., 2009), cerebrum (Hazell et al., 2009; Tang et al., 2014) and throughout the cerebellum, but most abundantly in Purkinje neurons (Hazell et al., 2009).

BLOOD–BRAIN BARRIER AND GLIAL CELLS

The endothelial cells of the CNS blood vasculature form the BBB, regulating the entry of substances into the brain. Providing support to the BBB are glial astrocyte cell projections, and together they regulate brain glucose uptake and energy metabolism. Estrogens are involved in the brain's ability to utilize glucose and a reduced capacity to take glucose up from the blood may contribute to neurodenerative diseases such as Alzheimer's disease (AD). Female rats treated with E2 treatment show an increase in glucose utilization while ovariectomized females have significantly reduced glucose utilization (Bishop and Simpkins, 1995). ERα and ERβ are both expressed in the endothelial cells of the BBB (Burek et al., 2014) and E2 has been shown to have differential effects on BBB permeability in female rats, with permeability increasing as circulating estrogens decrease with aging (Bake and Sohrabji, 2004).

Estrogens also affect glial cells via numerous mechanisms. Estrogen treatment can regulate astrocyte morphology in the hypothalamus and hippocampus (see McEwen and Alves, 1999 for review) which suggests a role for glial cells in synaptic plasticity. Estrogens can regulate Apolipoprotein E in microglia and astrocytes, and a deficit in Apolipoprotein E is linked to an insufficiency in hippocampal synaptic sprouting (Stone et al., 1998). Glial fibrillary acidic protein (GFAP) is also regulated by estradiol, resulting in either an increase or decrease in GFAP expression depending on the location in the brain. In the arcuate nucleus GFAP mRNA increases with E2 during proestrus (Kohama et al., 1995) while it is an absence of estrogens that results in an increase in GFAP expression in the hippocampus (Day et al., 1993).

ESTRADIOL AND BRAIN SEXUAL DIFFERENTIATION

Over a half-century ago Phoenix et al. (1959) suggested that sex differences in an animal's behavioral phenotype result from two types of hormone action, which were termed organizational (acting during a sensitive pre- or postnatal window) and activational (acting during the peripubertal period and adulthood). It is a universally accepted tenet that estradiol is one of the primary players in sexual differentiation of the rodent brain. Testosterone, secreted by the testes during fetal and neonatal life, is aromatized in the brain into E2, which acts to cause permanent changes resulting in a masculine pattern of brain development (see Figure 2.3). A two-stage model for the organizational effects of steroids has been proposed by Schulz and colleagues (2009) in which

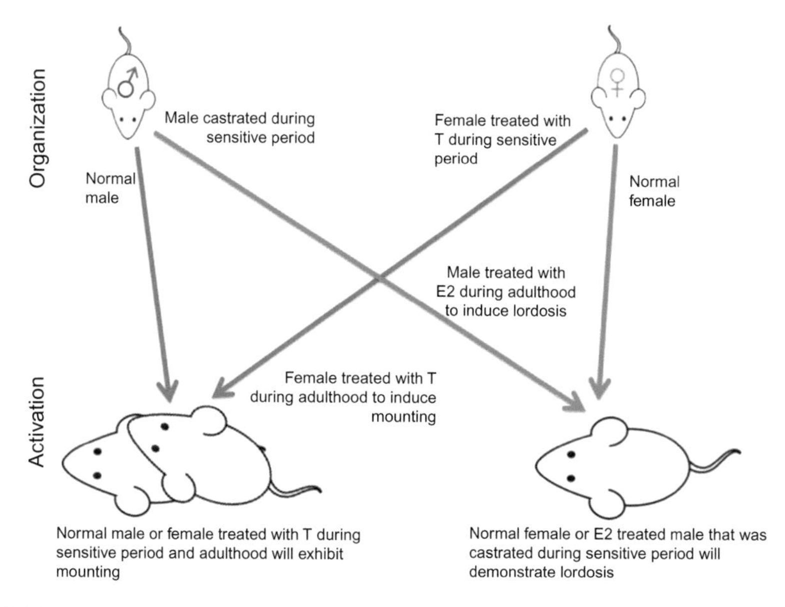

FIGURE 2.3 The hormonal regulation of rodent sexual differentiation illustrating organizational versus activational effects of estradiol on rodent mating behavior.

androgens need to be present during both the perinatal and the peripubertal periods in order to affect adult mating behavior. Organizational effects cannot be reversed later in life and are permanent. Then during the peripuberty period and adulthood, gonadal steroids act to contribute to sex differences in brain function and behavior.

RAPID ESTROGEN SIGNALING IN THE BRAIN

As with the synthesis of ovarian E2, brain-derived E2 is mediated by aromatase and, as mentioned earlier in this chapter, estrogens produced *de novo* may be the source of rapid estrogen signaling in the brain (see Figure 2.4). It is of great interest to determine if E2 synthesized in the brain influences circulating E2 levels, and while this is a possibility, much research has been conducted on the actions of brain-derived E2 acting exclusively in the brain.

A hypothesized mechanism for controlling the production and release of E2 from presynaptic terminals is synaptically localized aromatase (Remage-Healey et al., 2011; Srivastava et al., 2011). The process involves the presence of synaptically located aromatase regulating the rapid local synthesis of E2, followed by the activation of extranuclear ER coupled with intracellular signaling cascades via second messenger systems.

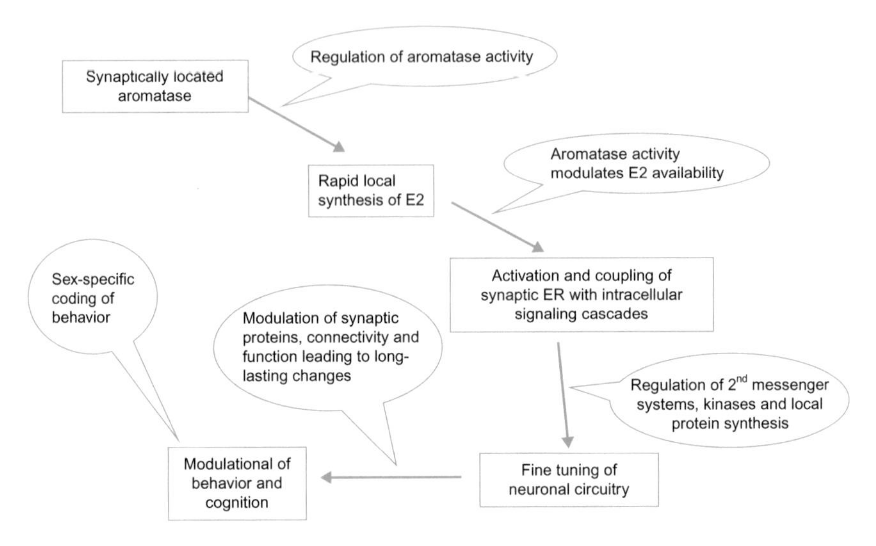

FIGURE 2.4 The source of fast E2 signaling in the brain results from E2 rapidly synthesized by aromatase located in the synapses. The presence of aromatase determines how much E2 will be available in the brain. *Adapted from Srivastava et al., 2011.*

As proposed by Srivastava and colleagues (2011) and others, this type of E2 signaling has been implicated in the "fine-tuning" of neuronal circuitry via alterations in the molecular and cellular machinery, ultimately resulting in the modulation of behavior.

One example where neuroestrogen has been extensively studied relates to the control of female reproduction via GnRH neuronal function. E2 treatment induces excitatory firing activity of GnRH cultured neurons cultured from embryonic primate nasal placode, and the proposed mechanism of action occurs through GPR30 extranuclear receptors (see Terasawa and Kenealy, 2012). Since this phenomenon results in a rapid change and not a long-term E2 action, two hypotheses of local E2 action in the hypothalamus have been proposed. The first is that the preovulatory surge of GnRH is augmented by local E2, and the second is that local E2 contributes to GnRH pulse frequency (Terasawa and Kenealy, 2012).

ESTROGEN-SENSITIVE BRAIN REGIONS

The following sections include examples of brain regions that are either sexually dimorphic or estrogen sensitive and respond to either gonadally derived or locally synthesized steroids.

Sexually Dimorphic Brain Regions

There are several areas of the brain where sexual dimorphisms have been described. The involvement of steroids in the sexual differentiation of the adult brain is readily evident in the sexually dimorphic nucleus of the preoptic area (SDN-POA) of the rodent hypothalamus, where there is a robust structural sex difference. The SDN-POA of males is substantially larger than that of untreated females (Gorski, 1984). Another POA sex difference that has been well described is in the medial preoptic area (MPOA). This sex difference is one of connectivity involving the synapsing of nonamygdaloid neurons in which male rats display more synapses on dendritic shafts and fewer on dendritic spines while females show the opposite pattern (Raisman and Field, 1973a), and this difference can be reversed by hormonal manipulation during the perinatal sensitive period (Raisman and Field, 1973b). In order for gonadally derived T to act on the brain during development, aromatase must be expressed in the appropriate neurons. Aromatase expression is at its highest in the sexually dimorphic regions of the hypothalamus, including the preoptic area, during the sensitive pre- and postnatal period (Roselli and Resko, 1993).

Hippocampus

The hippocampus is part of the limbic system and is implicated in explicit and spatial memory functions. It is important for the consolidation of short- and long-term memory formation, including declarative or verbal memory. Due to the presence of the appropriate steroid receptors, it is hormone sensitive and a major target for circulating adrenal and gonadal steroids. Intracellular ERs are located in the CA1, CA3, and dentate gyrus regions of the hippocampus. This region of the brain likely mediates the effects of estrogens on spatial memory performance as demonstrated by an improvement in delayed matching-to-place water maze trials in ovariectomized rats after acute and continuous estradiol treatment (Sandstrom and Williams, 2004). The hippocampus expresses both ERα and ERβ, and one proposed function of ERβ in the hippocampus is to regulate ERα-mediated transcription (Han et al., 2013).

In female rats, dendritic spine density in the CA1 region changes around ovulation when levels of ovarian estradiol are elevated. The ovarian cycle regulates synaptogenesis on hippocampal CA1 pyramidal neuron spines, and gonadectomy in females and males results in reduced CA1 dendritic spine density (Gould et al., 1990). However, estradiol treatment only rescues the effect in female, but not male rats (Leranth et al., 2004). Ultimately, increases or decreases in synapse density affect hippocampal dependent memory processes in female and male rats.

Data suggest that not only do ovarian estrogens alter the density of CA1 hippocampal pyramidal spines, but local synthesized estrogens are also involved (for review, see Spencer et al., 2008). Work by Vierk and colleagues (2014) demonstrates that hippocampally derived estradiol is the central facilitator in estrogen-induced synaptic plasticity in females. Estradiol synthesis in hippocampal cultures was inhibited via a variety of techniques showing that reduction of aromatase resulted in a significant reduction in the number of spine synapses in the CA1 neurons. This was further supported by the observation that spine density was reduced in female mice after systemic treatment with the aromatase inhibitor letrozole and, importantly, this effect was also seen in ovariectomized animals (as reviewed in Vierk et al., 2014).

Cerebellum

For a number of decades the cerebellum was considered to be insensitive to circulating steroids including estrogens; however, this view has changed. Estradiol has a diverse array of effects in the cerebellum including involvement in cell signaling during cerebellar development and modulation of synaptic transmission in adults (Hedges et al., 2012). The developing and mature cerebellum expresses both ERα and ERβ. Cerebellar granule cells express low levels of ERα (Ikeda and Nagai, 2006) while ERβ is expressed at higher levels in Purkinje and granule cells (Handa et al., 2012; Ikeda and Nagai, 2006). Recently it has been demonstrated that the cerebellum may be a site of local steroid synthesis and this may play a role in the function of cerebellar E2 to affect glutamtergic neurotransmission (Hedges et al., 2012). More specifically, this type of neurotransmission is reliant on E2 as was determined using an *in vivo* optical imaging of the parallel fiber-Purkinje neuron synapse (Hedges et al., 2012).

NEUROTRANSMITTER SYSTEMS

Estrogens also influence several neurotransmitter systems that project to the forebrain to regulate processes such as fine motor skills, motor activity, attention, and mood. Neuroactive steroids, such as estradiol, modulate classical synaptic transmission, including cholinergic, noradrenergic, dopaminergic and serotonergic synaptic transmission. This occurs by altering the responsiveness of postsynaptic receptors or the presynaptic release of neurotransmitter. We will briefly explore these four systems here.

Basal Forebrain Cholinergic System

The basal forebrain cholinergic system (neurons here express acetylcholine and cholinergic enzymes) is important for cognitive function, with neurons projecting to the cerebral cortex and hippocampus. This area is estrogen sensitive, and estrogen replacement induces choline acetylchotransferase (ChAT) activity in ovariectomized rats (McEwen et al., 1987). The hippocampus produces nerve growth factor (NGF), which is transported retrogradely to the basal forebrain cholinergic system. NGF is a proposed regulator of the cholinergic systems of the basal forebrain and ERα is co-localized with NGF receptors in cholinergic neurons in the basal forebrain (Toran-Allerand et al., 1992). Estrogen replacement in aged female rats increases expression of the NGF receptor trkA and ChAT (Singer et al., 1998), acting to promote neuronal survival.

Midbrain and Brainstem Serotonergic System

The serotonin system of the midbrain and brainstem Raphe nucleus is involved in a number of functions including aggression, mood, sleep, feeding, growth, reproduction, and cognition, with serotonergic neurons projecting to forebrain regions such as hypothalamus, hippocampus, and cortex. A sex dimorphism in the serotonin system of the rat is established early in the postnatal period (Carlsson and Carlsson, 1988) with females having higher serotonin levels in several areas as compared to males and ovarian steroids affect serotonin function in a sexually dimorphic manner. Estrogen treatment increases expression of tryptophan hydroxylase (TPH), the rate-limiting enzyme in the production of serotonin, and decreases the expression of SERT, the serotonin transporter (Bethea et al., 1999). Evidence supports a greater role for ERβ than ERα in facilitating estrogen action in the dorsal Raphe nucleus. This is supported by data that shows that over 90% of ERβ-immunoreactive neurons in the Raphe nucleus are co-localized with TPH while only 23% of ERα-immunoreactive neurons are (Nomura et al., 2005).

Catecholaminergic Neurons

Catecholaminergic neurons are those that contain either the neurotransmitters dopamine (dopaminergic system) or norepinephrine (noradrenergic system). These neurons are estrogen-sensitive, and brainstem catecholaminergic neurons contain small numbers of intracellular ER (McEwen and Alves, 1999). The enzyme tyrosine hydroxylase is the rate-limiting step in the conversion of L-tyrosine to L-DOPA in the synthesis of dopamine and then norepinephrine, and estradiol regulates tyrosine hydroxylase gene expression in catecholaminergic neurons. Estrogen replacement after gonadectomy alters tyrosine hydroxylase mRNA expression in a time-dependent manner (Liaw et al., 1992). More specifically, both ERα and early gene expression (Jennes et al., 1992) show cyclic patterns in A1 and A2 noradrenergic neurons in both different animal models (Simonian et al., 1998).

The midbrain dopaminergic system includes projections to the nucleus accumbens and corpus striatum and is important for motor, cognitive and affective behavior. In the corticolimbic and striatal targets of midbrain dopaminergic neurons, estrogen can modulate the synthesis, release and metabolism of dopamine; however, the mechanism of this action has yet to be fully characterized (Creutz and Kritzer, 2002). The effects of estrogen on dopaminergic neurons are likely through both intracellular and extranuclear estrogen receptors (Küppers et al., 2000); however, elevated levels of ERβ-immunoreactivity are present in the midbrain dopaminergic system. In the brains from hormonally intact male and female rats, the distribution of ERβ-immunoreactive cells was confined to dopaminergic and nondopaminergic neurons within the dorsal ventral tegmental area, the substantia nigra pars lateralis, the parabrachial pigmented nucleus, the retrorubral fields, and linear midline nuclei (Creutz and Kritzer, 2002). As with noradrenergic neurons, estrogen also stimulates the expression of tyrosine hydroxylase in dopaminergic neurons. Finally, in striatum, the primary input for the basal ganglia, is also estrogen sensitive, and both ERα and ERβ mRNA are expressed in the developing and adult mouse striatum (Küppers and Beyer, 1999). Several studies have shown that, in striatum, the depolarization-induced release of dopamine decreases after ovariectomy and increases after estradiol replacement (see McEwen and Alves, 1999 for review).

ESTROGENS, AGING, AND NEUROPROTECTION

Estrogens influence brain morphology and functioning throughout an animal's lifespan, and especially during development and aging, and there may be changes that occur during aging that alter how the brain responds to estrogens. As will be explored in more detail in the subsequent chapters in this text, estradiol has been implicated in a critical role in protection against cognitive decline related to aging, stroke, and neurodegenerative diseases such as AD, and both gonadally derived and locally synthesized estrogens are involved. This has been explored in humans where estrogen replacement therapy in postmenopausal women combats age-related cognitive decline and possibly provides protective effects against AD, and in various animal models (reviewed in Norbury et al., 2003). Estrogen treatment protects hippocampus-dependent memory in the aging brain (Bimonte-Nelson et al., 2010) and estrogen replacement can affect brain neurochemistry and function, neurotransmitter systems, macroscopic brain structure, and metabolism (reviewed in Norbury et al., 2003).

A recently described potential mechanism involved in age-related changes in neuronal responsiveness to estradiol may involve ERβ, and ERβ protein-protein interactions. ERβ is the predominant ER in the hippocampus and is responsible for functions that range from gene transcription to synaptic transmission. Research by Mott and colleagues (2014) has shown that in the ventral hippocampus there is an intrinsic change in ERβ function, and that this receptor interacts with Annexin proteins (involved in a variety of cellular and physiological processes) differentially depending on age.

Protection against damage produced by ischemia in experimental models of stroke can be conferred by estrogen treatment. In addition, animal models of spinal cord injury also suggest that estradiol acts in a neuroprotective manner as well as showing anti-apoptotic and anti-inflammatory properties (Elkabes and Nicot, 2014). One technique to validate how estradiol may be involved in neuroprotection after ischemia is to use estradiol conjugated to bovine serum albumin (E2-BSA). This conjugate is cell impermeable and can only interact with extranuclear ERs, such as GPR30 resulting in activation of kinase signaling pathways. Neuroprotection after cerebral ischemia involving rapid estrogen signaling regulated by GPR30 was determined by knockdown of GPR30 protein with E2-BSA treatment *in vivo* in hippocampal CA1 neurons (Tang et al., 2014). Estradiol working through this mechanism can result in activation of prosurvival kinases that may facilitate neuroprotection.

To further support a role for glia in estrogen-related neuroprotection, animal models of global cerebral ischemia (GCI) can be used. Ovariectomized female rats exposed to brief GCI had increased levels of aromatase in GFAP-positive astrocytes and hippocampal CA1 neurons, while nonischemic females had increased aromatase staining in hippocampal CA1, CA3 and dentate gyrus (Zhang et al., 2014). This result was accompanied by an increase in local estradiol synthesis and blocked by central aromatase antisense oligonucleotide administration (Zhang et al., 2014).

Finally, an alternative to E2 treatment is the possible use of SERMS, or selective estrogen receptor modulators, to exert estradiol-like neuroprotective actions in the brain (DonCarlos et al., 2009). SERMS can stimulate the neuroprotective mechanisms and reduce neural damage in animal models of ischemia, neurodegenerative disease, and cognitive impairment. As with steroids, SERMS bind to intracellular ERs to

ESTROGEN EFFECTS ON TRAUMATIC BRAIN INJURY

facilitate transcription. SERMs include tamoxifen, and may act as ER agonists in the brain and as antagonists in others tissues, such as cancerous tumors in breast tissue.

ADDITIONAL PHYSIOLOGICAL EFFECTS OF ESTROGENS ON THE BODY

In addition to their robust actions on the female reproductive system and sexual behavior, estrogens also alter several other physiological systems. They play a role in calcium metabolism, resulting in increased bone production. The increase in estrogen secretion during sexual development and puberty results in an increase in bone mineral density and the closing of the epiphyseal plate in long bones due to an increase in apoptosis of chondrocytes (Zhong et al., 2011). In women, more bone is made in the presence of high concentration of estrogens (Frenkel et al., 2010), and during menopause, when estrogen concentrations decrease, osteoporosis can result. Another aspect of E2 action that can affect overall body physiology is with body fluid regulation. E2 lowers the osmotic threshold for vasopressin (AVP) release resulting in an increased plasma volume (Stachenfeld, 2008). A recent finding suggests that in female rats brown adipose tissue thermogenesis is regulated via estradiol acting through ERα. Estradiol does this by inhibiting AMP-activated protein kinase in the ventromedial nucleus of the hypothalamus (Martínez de Morentin et al., 2014). This demonstrates the variety of actions that estrogens can exhibit in the body and brain.

CONCLUSIONS AND NEW DIRECTIONS

An ever-growing body of research is altering our views of how hormones act in the brain to regulate behavior. Exciting findings have challenged the traditional dogma of how estrogens act in the brain. The classical understanding of estrogens acting through genomic mechanisms to influence physiology and behavior has expanded to include extranuclear receptors and site-specific estradiol synthesis. Recent research also suggests that the rapid signaling of estrogens occurs via membrane-bound receptors, supporting the hypothesis that steroids can be synthesized *de novo* in tissues such as the brain without requiring gonadally derived steroids as precursors. Described in this chapter are examples that illustrate the versatility of estrogens whose functions extend beyond just regulating female reproductive physiology and behavior. Estrogens have been justifiably described by McEwen and

Alves (1999) as "multipurpose messengers" whose actions are extended throughout many brain regions and across the lifespan. As our understanding of the mechanisms of how estrogens act to influence neuronal functioning grows, new roles for estrogens have emerged. Estrogens play a pivotal role as neurotherapeutic agents resulting in neuroprotection during aging and after brain injury. New challenges include developing therapies that utilize the mechanisms of action of estrogens in neuroprotection while minimizing the risks associated with estrogen replacement therapy.

References

Bake, S., Sohrabji, F., 2004. 17β-Estradiol differentially regulates blood–brain-barrier permeability in young and aging female rats. Endocrinology. 145, 5471–5475.

Beach, F.A., 1948. Hormones and Behavior. Paul B. Hoeber, New York.

Beato, M., Klug, J., 2000. Steroid hormone receptors: An update. Hum. Reprod. Update. 6, 225–236.

Bethea, C.L., Pecins-Thompson, M., Schutzer, W.E., Gundlah, C., Lu, Z.N., 1999. Ovarian steroids and serotonin neural function. Mol. Neurobiol. 18, 87–123.

Bishop, J., Simpkins, J.W., 1995. Estradiol enhances brain glucose uptake in ovariectomized rats. Brain Res. Bull. 36, 315–320.

Bimonte-Nelson, H.A., Acosta, J.I., Talboom, J.S., 2010. Neuroscientists as cartographers: mapping the crossroads of gonadal hormones, memory and age using animal models. Molecules. 15, 6050–6105.

Boling, F.L., Blandau, R.J., 1939. The estrogen-progesterone induction of mating responses in the spayed female rat. Endocrinology. 25, 359–364.

Brailoiu, E., Dun, S.L., Brailoiu, G.C., Mizuo, K., Sklar, L.A., Oprea, T.I., et al., 2007. Distribution and characterization of estrogen receptor G protein-coupled receptor 30 in the rat central nervous system. J. Endocrinol. 193, 311–321.

Burek, M., Steinberg, K., Förster, C.Y., 2014. Mechanisms of transcriptional activation of the mouse cluadin-5 promoter by estrogen receptor alpha and beta. Mol. Cell. Endocrinol. 392, 144–151.

Caldwell, J.D., Jirikowski, G.F., 2014. Sex hormone binding globulin and corticosteroid binding globulin as major effectors of steroid action. Steroids. 81, 13–16.

Carlsson, M., Carlsson, A., 1988. A regional study of sex differences in rat brain serotonin. Prog. Neuropsychopharmacol. Biol. Psychiatry. 12, 53–61.

Creutz, L.M., Kritzer, M.F., 2002. Estrogen receptor-beta immunoreactivity in the midbrain of adult rats: regional, subregional, and cellular localization in the A10, A9, and A8 dopamine cell groups. J. Comp. Neurol. 446, 288–300.

Day, J.R., Laping, N.J., Lampert-Etchells, M., Brown, S.A., O'Callaghan, J.P., McNeill, T.H., et al., 1993. Gonadal steroids regulate the expression of glial fibrillary acidic protein in the adult male rat hippocampus. Neuroscience. 55, 435–443.

DonCarlos, L.L., Azcoitia, I., Garcia-Segura, L.M., 2009. Neuroprotective actions of selective estrogen receptor modulators. Psychoneuroendocrinology. 34, S113–S122.

Elkabes, S., Nicot, A.B., 2014. Sex steroids and neuroprotection in spinal cord injury: a review of preclinical investigations. Exp. Neurol. 259, 28–37.

Frenkel, B., Hong, A., Baniwal, S.K., Coetzee, G.A., Ohlsson, C., Khalid, O., et al., 2010. Regulation of adult bone turnover by sex steroids. J. Cell. Physiol. 224, 305–310.

Gorski, R., 1984. Critical role for the medial preoptic area in the sexual differentiation of the brain. Prog. Brain Res. 61, 129–146.

Gould, E., Woodley, C., Frankfurt, M., McEwen, B.S., 1990. Gonadal steroids regulate dendritic spine density in hippocampal pyramidal cells in adulthood. J. Neurosci. 10, 1286–1291.

Han, X., Aenile, K.K., Bean, L.A., Rani, A., Semple-Rowland, S.L., Kumar, A., et al., 2013. Role of estrogen receptor α and β in preserving hippocampal function during aging. J. Neurosci. 33, 2671–2683.

Handa, R.J., Ogawa, S., Wang, J.M., Herbison, A.E., 2012. Roles for oestrogen receptor beta in adult brain function. J. Neuroendocrinol. 24, 160–173.

Hazell, G.G., Yao, S.T., Roper, J.A., Prossnitz, E.R., O'Carroll, A.M., Lolait, S.J., 2009. Localisation of GPR30, a novel G protein-coupled oestrogen receptor, suggests multiple functions in rodent brain and peripheral tissues. J. Endocrinol. 202, 223–236.

Hedges, V.L., Ebner, T.J., Meisel, R.L., Mermelstein, P.G., 2012. The cerebellum as a target for estrogen action. Front. Neuroendocrinol. 33, 403–411.

Harada, N., Utsumi, T., Takagi, Y., 1993. Tissue-specific expression of the human aromatase cytochrome P-450 gene by alternate use of multiple exons 1 and promoters, and switching of tissue-spe exons 1 in carcinogenesis. Proc. Natl. Acad. Sci. USA. 90, 11312–11316.

Ikeda, Y., Nagai, A., 2006. Differential expression of the estrogen receptors alpha and beta during postnatal development of the rat cerebellum. Brain Res. 1083, 39–49.

Jennes, L., Jennes, M.E., Purvis, C., Nees, M., 1992. c-fos Expression in noradrenergic A2 neurons of the rat during the estrous cycle and after steroid hormone treatments. Brain Res. 586, 171–175.

Kohama, S.G., Goss, J.R., Finch, C.E., McNeill, T.H., 1995. Increases in glial fibrillary acidic protein in the aging female mouse brain. Neurobiol. Aging. 16, 59–67.

Kuiper, G.G.J.M., Shughrue, P.J., Merchenthaler, I., Gustafsson, J.-A., 1998. The estrogen receptor β subtype: a novel mediator of estrogen action in neurodendocrine systems. Front. Neurocrinol. 19, 235–286.

Küppers, E., Beyer, C., 1999. Expression of estrogen receptor-alpha and beta mRNA in the developing and adult mouse striatum. Neurosci. Lett. 276, 95–98.

Küppers, E., Ivanova, T., Karolczak, M., Beyer, C., 2000. Estrogen: A multifunctional messenger for nigrostriatal dopaminergic neurons. J. Neurocytol. 29, 375–385.

Leranth, C., Hajszan, T., MacLusky, N.J., 2004. Androgens increase spine synapse density in the CA1 hippocampal subfield of ovariectomized female rats. J. Neurosci. 24, 495–499.

Liaw, J.J., He, J.R., Hartman, R.D., Barraclough, C.A., 1992. Changes in tyrosine hydroxylase mRNA levels in medullary A1 and A2 neurons and locus coeruleus following castration and oestrogen replacement in rats. Brain Res. Mol. Brain Res. 13, 213–238.

Lovejoy, D.A., 2005. Neuroendocrinology: An Integrated Approach. John Wiley & Sons, Ltd., West Sussex, England.

Martínez de Morentin, P.B., González-García, I., Martins, L., Lage, R., Fernández-Mallo, D., Martínez-Sánchez, N., et al., 2014. Estradiol regulates brown adipose tissue thermogenesis via hypothalamic AMPK. Cell Metabolism. 20 (1), 41–53.

McEwen, B.S., Alves, S.H., 1999. Estrogen actions in the central nervous system. Endocr. Rev. 20, 279–307.

McEwen, B.S., Luine, V., Fischette, C., 1987. Developmental actions of hormones: from receptors to function. In: Easter, S., Barald, K., Carlson, B. (Eds.), From Message to Mind. Sinauer Associates, Sutherland, MA, pp. 272–287.

Mott, N.N., Pinceti, E., Rao, Y.S., Przybycien-Szymanska, M.M., Prins, S.A., Shults, C.L., et al., 2014. Age-dependent effects of 17β-estradiol on the dynamics of estrogen receptor β (ERβ) protein-protein interactions in the ventral hippocampus. Mol. Cell. Proteomics. 13, 760–779.

Nomura, M., Akama, K.T., Alves, S.E., Korach, K.S., Gustafsson, J.-A., Pfaff, D.W., et al., 2005. Differential distribution of estrogen receptor (ER)-α and ER-β in the midbrain

raphe nuclei and periaqueductal gray in male mouse: predominant role of ER-β in midbrain serotonergic systems. Neuroscience. 130, 445–456.
Norbury, R., Cutter, W.J., Compton, J., Robertson, D.M., Craig, M., Whitehead, M., et al., 2003. The neuroprotective effects of estrogen on the aging brain. Exp. Gerontol. 38, 109–117.
Pérez, S.E., Chen, E.-Y., Mufson, E.J., 2003. Distribution of estrogen receptor alpha and beta immunoreactive profiles in the postnatal rat brain. Dev. Brain Res. 145, 117–139.
Raisman, G., Field, P.M., 1973a. Sexual dimorphism in the preoptic area of the rat. Science. 173, 731–733.
Raisman, G., Field, P.M., 1973b. Sexual dimorphism in the neutrophil of the preoptic area of the rat and its dependence on neonatal androgen. Brain Res. 54, 1–29.
Remage-Healey, L., Dong, S., Maidment, N.T., Schlinger, B.A., 2011. Presynaptic control of rapid estrogen fluctuations in the songbird auditory forebrain. J. Neurosci. 31, 100034–110038.
Roselli, C.E., Resko, A.A., 1993. Aromatase activity in the rat brain: hormonal regulation and sex differences. J. Steroid Biochem. Mol. Biol. 44, 499–508.
Phoenix, C.H., Goy, R.W., Gerall, A.A., Young, W.C., 1959. Organizing action of prenatally administered testosterone propionate on the tissues mediating behavior in the female guinea pig. Endocrinology. 65, 369–382.
Sandstrom, N.J., Williams, C.L., 2004. Spatial memory retention is enhanced by acute and continuous estradiol replacement. Horm. Behav. 45, 128–135.
Schulz, K.M., Molendo-Figueira, H.A., Sisk, C.L., 2009. Back to the future: The organizational-activational hypothesis adapted to puberty and adolescence. Horm. Behav. 55, 597–604.
Shughrue, P.J., Komm, B., Merchenthaler, I., 1996. The distribution of estrogen receptor-β mRNA in the rat hypothalamus. Steroids. 61, 678–681.
Simonian, S.X., Delaleu, B., Caraty, A., Herbison, A.E., 1998. Estrogen receptor expression in brainstem noradrenergic neurons of the sheep. Neuroendocrinology. 67, 392–402.
Singer, C.A., McMillan, P.J., Dobie, D.J., Dorsa, D.M., 1998. Effects of estrogen replacement on choline acetyltransferase and trkA mRNA expression in the basal forebrain of aged rats. Brain Res. 789, 343–346.
Spencer, J.L., Waters, E.M., Romeo, R.D., Wood, G.E., Milner, T.A., McEwen, B.S., 2008. Uncovering the mechanisms of estrogen effects on hippocampal function. Front. Neuroendocrinol. 29, 219–237.
Srivastava, D.P., Waters, E.M., Mermelstein, P.G., Kramár, E.A., Shors, T.J., Liu, F., 2011. Rapid estrogen signaling in the brain: implications for the fine-tuning of neuronal circuitry. J. Neurosci. 31, 16056–16063.
Stachenfeld, N.S., 2008. Sex hormone effects on body fluid regulation. Exerc. Sport Sci. Rev. 36, 152–159.
Stone, D.J., Rozovsky, I., Morgan, T.E., Anderson, C.P., Finch, C.E., 1998. Increased synaptic sprouting in response to estrogen via an apolipoprotein E-dependent mechanism: implications for Alzheimer's disease. J. Neurosci. 18, 3180–3185.
Tang, H., Zhang, Q., Yang, L., Dong, Y., Khan, M., Yang, F., et al., 2014. Reprint of GPR30 mediates estrogen rapid signaling and neuroprotection. Mol. Cell Endocrinol. 389, 92–98.
Terasawa, E., Kenealy, B.P., 2012. Neuroestrogen, rapid action of estradiol, and GnRH neurons. Front. Neuroendocrinol. 33, 364–375.
Tetel, M.J., 2009. Modulation of steroid action in the central and peripheral nervous systems by nuclear receptor coactivators. Psychoneuroendocrinology. 34, S9–S19.
Toran-Allerand, C.D., Miranda, R.C., Bentham, W.D.L., 1992. Estrogen receptors colocalize with low-affinity nerve growth factor receptors in cholinergic neurons of the basal forebrain. Proc. Natl. Acad. Sci. USA 89, 4668–4672.

Vierk, R., Brandt, N., Rune, G.M., 2014. Hippocampal estradiol synthesis and its significance for hippocampal synaptic stability in male and female animals. Neuroscience. 274, 24–32.

Zhang, Q.-G., Wang, R., Tang, H., Dong, Y., Chan, A., Sareddy, G.R., et al., 2014. Brain-derived estrogen exerts anti-inflammatory and neuroprotective actions in the rat hippocampus. Mol. Cell. Endocrinol. 389, 84–91.

Zhong, M., Carney, D.H., Boyan, B.D., Schawartz, Z., 2011. 17β-estradiol regulates rat growth plate chondrocyte apoptosis through a mitochondrial pathway not involving nitric oxide or MAPKs. Endocrinology. 152, 82–92.

CHAPTER

3

Induction of Estrogen Response Following Injury

Suzanne R. Burstein[1] *and Kelli A. Duncan*[2]

[1]Weill Cornell Medical College, New York, New York, USA [2]Biology and Neuroscience and Behavior, Vassar College, Poughkeepsie, New York, USA

INTRODUCTION

In 1873, Camillo Golgi developed a silver staining technique (later named the Golgi method) that allowed him to visualize nervous tissue using a light microscope and was one step closer to identifying the various types of cells within the nervous system (DeFelipe, 2013). Around the same time, Rudolf Virchow had just discovered a class of cells within the nervous system and described them as being the glue (neuroglia) that held the neuronal elements together (Garcia-Segura et al., 1996). Years later, work by Ramón y Cajal and Wilhelm His would further characterize the specific cells of the nervous system and lay the foundation for the study of their anatomy and function (DeFelipe, 2013). We now know that the brains of all species are composed primarily of two broad classes of cells: neurons and glia. Neurons serve as the basic building blocks of the nervous system and are traditionally thought to be the most important cells. Despite the prevalence and importance of neurons in the nervous system, glial cells are more abundant and are just as important as neurons for normal function.

It is estimated that there are between 10 and 50 times more glial cells than neurons in the human brain (Snell, 2009). In the past, their only described function was to provide support and protection for neurons as well as to regulate cell migration during development. There are several different types of glial cells: astrocytes, oligodendrocytes, microglia, ependymal cells, radial glial, satellite cells and Schwann cells. Years after Golgi identified glial cells in the CNS,

K.A. Duncan (Ed): Estrogen Effects on Traumatic Brain Injury.
DOI: http://dx.doi.org/10.1016/B978-0-12-801479-0.00003-6

researchers have now identified that astroglia/astrocytes, radial glia, and microglia appear to be both the targets and mediators of innate mechanisms of neuroprotection (specifically estrogen mediated neuroprotection), and are not simply the "glue" that holds the nervous system together (Arevalo et al., 2013; Burda and Sofroniew, 2014). Within this chapter, we will discuss the role of glial cells in local steroid synthesis and neuroprotection near injury sites. We will also examine the connection between glial aromatization and neuroimmune responses to injury.

REACTIVE GLIOSIS FOLLOWING INJURY

Injury to the brain causes a series of pathological signaling cascades that can be divided into two phases: 1) the immediate mechanical damage, and 2) the inflammatory response (Ghirnikar et al., 1998; Marciano et al., 2002). The neuroinflammatory response to injury can be both neurotoxic and neuroprotective and greatly influences the outcome of the injury (Lenzlinger et al., 2001). Following injury, a complex phenomenon involving the activation and proliferation of glial cells and the migration of these cells towards the sites of injury occurs (Graeber and Streit, 2010; Lull and Block, 2010; Sofroniew, 2009), termed *reactive gliosis*. Reactive gliosis regulates the early repair and protection mechanisms following damage. These activated glial cells release chemokines and cytokines locally that regulate the inflammatory processes that mediate edema and swelling (Elkabes et al., 1996).

Microglia are the smallest of the glial cells and serve as the representative of the immune system in the brain (Figure 3.1). Microglia

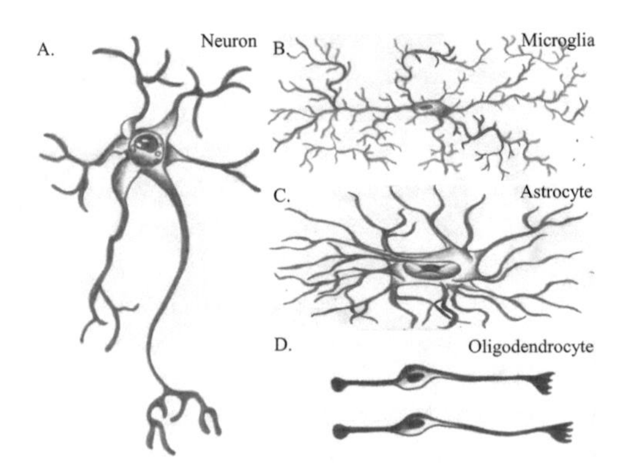

FIGURE 3.1 Representative images of glial cell subtypes.

protect the brain from invading microorganisms and are thought to be the resident macrophages of the CNS (Prinz and Mildner, 2011). Following injury, microglia are the first to respond and perform a number of tasks to help protect the damaged neurons (Robel et al., 2011). This interaction or "crosstalk" with neurons is part of a multi-staged response that provides neuroprotection to neurons following injury. Initially, the neuron sends out a rescue signal that activates microglia from their dormant state (Streit et al., 2008). Attracted by endogenous and other chemotactic factors, microglial cells migrate toward the site of brain injury. They engulf substances inadvertently produced from the infiltration of polymorphonuclear cells and other blood cells, which offers neuroprotection to the surrounding tissue. Next, these activated microglial cells release neurotrophic factors that are important in nerve cell repair and neuroprotection. These factors include trophic molecules such as transforming growth factor-β1 (TGF-β1). TGFβs are anti-inflammatory and neuroprotective cytokines, which limit neuroinflammation following injury (Cekanaviciute et al., 2014). While this activation of cytokines and chemokines by microglia is neuroprotective, chronic activation can lead to a breakdown of the blood–brain barrier and the production of reactive oxygen species (ROS) that can lead to increased brain damage (Michels et al., 2014).

Radial glial cells (Figure 3.1) are the primary progenitor cells capable of generating neurons, astrocytes, and oligodendrocytes (Malatesta and Gotz, 2013; Sild and Ruthazer, 2011). During development, newborn neurons use radial glia as scaffolds, traveling along the radial glial fibers in order to reach their final destinations (Sild and Ruthazer, 2011). Following injury, radial glial cells are upregulated around the injury site in both the brain and spinal cord of vertebrates (Peterson et al., 2004; Shibuya et al., 2002; Wu et al., 2005; Zupanc and Zupanc, 2006). Presumably, these cells are being "regenerated" to promote neurogenesis to replace the injured and damaged cells (Malatesta and Gotz, 2013; Wu et al., 2005). The effect of radial glia on neuroprotection, specifically neurogenesis, will be discussed in more detail later (see Chapter 5). The term "radial glia" refers to the morphological characteristics of these cells: namely, their radial processes and their similarity to another glial cell, astrocytes.

Astrocytes are the most numerous nonneuronal cells in the central nervous system and make up about 50% of the human brain volume (Chen and Swanson, 2003; Tower and Young, 1973). Astrocytes, like radial glia, are primarily attributed to providing structure for neurons. Like microglia, they also have the ability to rid the CNS of debris and dead cells via phagocytosis. However, we also know that they play a very important role following injury (Acaz-Fonseca et al., 2014; Episcopo et al., 2013; Sohrabji, 2014; Yuan and He, 2013). Astrocytes and microglia are involved in the complex interaction of both pro- and anti-inflammatory processes that can lead to both neuroprotection and neurodegeneration (Burda and Sofroniew, 2014). The astrocytic response is the "slow burn" of the reactive glial response; it is slower and longer lasting than the microglial response following injury (Streit et al., 2005). Astrocyte effects on brain injury are divided into those with immediate influence on cell survival and those with long-term or delayed effects that influence later recovery and function (Chen and Swanson, 2003). Reactive astrocytes suffer a continuum of progressive and multiple modifications in gene expression and cellular morphology that depend on the cellular context and the severity of the insult (Sofroniew, 2009; Sofroniew and Vinters, 2010). These modifications, which involve gain and loss of cellular functions, may have both beneficial and detrimental effects on the surrounding tissue. Immediately following injury, astrocytes are involved in decreasing glutamate toxicity and release a number of metabolites. Furthermore, astrocytes are more capable of scavenging ROS than neurons, suggesting that oxidant-scavenging mechanisms in astrocytes may function to support neuronal survival (Chen and Swanson, 2003). The second stage of astrocytic function following injury includes the effect on secondary injury repair. Like microglia, astrocytes also release cytokines, chemokines, and trophic factors following injury. The next chapter will delve deeper into the effect of glial astrocytic mediated neuroprotection and recovery following injury.

In summary, acute reactive gliosis and acute inflammation are considered adaptive responses that contribute to minimizing neuronal damage. This is done via the release of chemokines, cytokines, and neurotrophic factors to minimize damage and promote neurogenesis and survival. However, chronic or exacerbated gliosis and inflammation can enhance neuronal damage and amplify the neurodegenerative process (Arevalo et al., 2013; Chen and Swanson, 2003; Graeber and Streit, 2010; Lull and Block, 2010; Sofroniew, 2009). An interesting example of this complex relationship can be observed here. Following injury astrocytes release inducible nitric oxide synthase (iNOS) (Endoh et al., 1993). Nitric oxide is an ROS that can contribute to neuronal cell death by potentiating glutamate excitotoxicity (Chen and Swanson, 2003; Hewett et al., 1994) and by several other mechanisms (Chen and Swanson, 2003; Dawson and Dawson, 1998). However, when researchers decreased iNOS levels in mice, the mice exhibited smaller injury sizes (Chen and Swanson, 2003; Iadecola et al., 1997). Paradoxically, another substance released from astrocytes is steroid hormone. When aromatase (enzyme that converts androgens to estrogen) expression is blocked in these cells following injury, the resulting size of injury is much larger than control animals (Wynne and

Saldanha, 2004); this suggests that there exists a dichotomy in the role of astrocytes following injury.

NEW ROLES FOR GLIAL CELLS AS STEROID SYNTHESIZING CELLS

For quite some time, we have known that glial cells are both able to synthesize and respond to steroid hormones and may provide a link between the endocrine and nervous systems (Garcia-Segura et al., 1996; Johann and Beyer, 2013). *In vivo*, constitutive expression of aromatase is readily detectable in radial glia in the uninjured teleost brain (Xing et al., 2014; Zhang et al., 2014). These data suggest that radial glial cells have the potential for *de novo* steroid synthesis from cholesterol. These observations raise the possibility that constitutive steroidogenesis may provide the evolutionary framework for neuroprotection following injury (Saldanha et al., 2009). *In vitro* studies suggest that glial cells have all of the necessary enzymes to participate in the neurosteroid synthesis of testosterone, estrogen and progesterone (Garcia-Segura et al., 1996). Astrocytes also express receptors for all steroid receptors including glucocorticoids, estrogens, androgens, and progesterone receptors (Garcia-Segura et al., 1996). There is convincing evidence that astrocytes from different brain regions and in different phases of life contain all relevant types of receptors for both steroid hormones, although it is not fully clear whether these receptors are always expressed constitutively or rather on demand when astrocytes become activated under pathological conditions (Johann and Beyer, 2013).

Aromatase is constitutively expressed in neurons in regions of the uninjured brain. Following traumatic brain injury, both mammals and songbirds exhibit an increase in aromatase in reactive astrocytes located near the injury site (Azcoitia et al., 2001; Garcia-Segura et al., 1999b; Peterson et al., 2001; Wynne and Saldanha, 2004; Wynne et al., 2008b). This upregulation of aromatase expression also occurs in response to excitotoxic stimuli (Azcoitia et al., 2001; Garcia-Segura et al., 1999a). An extensive review of the role of aromatase and injury will be presented in the next chapter. Briefly, induction of glial aromatase decreases injury size and promotes neuroprotection. Estrogen receptor alpha and beta have both been shown to be upregulated following various types of neuronal insult (Garcia-Segura et al., 1999a; Johann and Beyer, 2013) and estrogen receptor-alpha deletion in astrocytes, but not in neurons, mediates the neuroprotective estrogenic effects. These data suggest that estrogens play a key role in neuroprotection following injury; however, how these reactive glial cells become agents of local steroid production remains elusive.

INDUCTION OF GLIAL AROMATASE FOLLOWING INJURY

The induction of glial aromatase after traumatic brain injury occurs in both mammals and in birds, with some differences that have been noted. In zebra finches, the increase in aromatase after injury occurs in radial glia (Peterson et al., 2004) in addition to reactive astrocytes, and happens rapidly and robustly after the injury. Also, upregulation of aromatase prevents a wave of secondary degeneration following traumatic brain injury in the zebra finch (Wynne et al., 2008a), which is lost when aromatase is inhibited with fadrozole before injury. The passerine model highlights the important role of the local induction of aromatase mechanism in neuroprotection. The protection conferred by aromatization has been shown in several models. Aromatase knockout mice lack estrogen and display increased apoptosis (Hill et al., 2009). After injury, estrogen can enhance neurogenesis and cytogenesis (Peterson et al., 2007; Walters et al., 2011). The specific mechanisms by which estrogen synthesized by glial cells affects neuronal viability and function are not yet fully elucidated, and will be discussed further in a later chapter. One growth factor, bone-morphogenetic-protein 2 (BMP2), is upregulated in microglia near the injury site in an aromatase-dependent manner, and is likely involved in the neuroprotection that ensues (Walters and Saldanha, 2008).

Injury induces expression, specifically in glia, of the 1a splice variant of aromatase in the zebra finch. This variant is normally expressed constitutively in neurons in some brain regions, and is different from the 1b variant expressed in the ovary (Ramachandran et al., 1999). Thus, the induction of glial aromatase after injury must be under different transcriptional control mechanisms than when expressed constitutively in neurons (Wynne et al., 2008a). In the songbird, cells in the song circuit often die without any induction of glial aromatase (Alvarez-Buylla and Kirn, 1997; Smith et al., 1997), suggesting that glial aromatase upregulation could be a response to the inflammation after injury rather than to apoptosis itself. A full understanding of the mechanism of aromatase induction would provide insight into the molecular events that follow traumatic brain injury, and can identify the pathways that contribute to neuroprotection via locally synthesized estrogen.

Neuroinflammation Induces Glial Aromatase Independent of Injury

Many studies have shown that cytokines, particularly interleukin-6 (IL-6) and interleukin 1β (IL-1β), regulate aromatase in cancer cells (Purohit et al., 1995; Purohit et al., 2005; Reed et al., 1993; Singh et al.,

1997). In breast tumors, both aromatase and pro-inflammatory cytokines are elevated (Bukulmez et al., 2008), and it is thought that cytokines regulate cancer cell proliferation via steroid-catalyzing enzymes such as aromatase (Honma et al., 2002). In endometriosis, an estrogen-dependent inflammatory disease, aromatase expression is upregulated in patient biopsies (Bukulmez et al., 2008), and estrogen contributes to an increase in inflammation and disease progression that can be treated with aromatase inhibition (Ferrero et al., 2014). Similarly, in endometrial carcinoma, a positive feedback loop causes estrogen produced in cancer cells to increase IL-6, increasing aromatization in stromal cells and enhancing estrogen synthesis in the tumor environment (Che et al., 2014). Taken together, the evidence from other fields identifies a strong interplay between inflammation and aromatase/estrogen induction that may also occur in the brain.

Accordingly, in traumatic brain injury models, inflammation also induces aromatase expression. Following a mechanical traumatic brain injury in the zebra finch in the entopallium, an area in which constitutive neuronal aromatase is not expressed, IL-6 and IL-1β are upregulated in the area surrounding the injury prior to aromatase upregulation. When an inflammatory response is evoked without mechanical injury, using the toxin phytohemagglutinin, glial aromatase expression is induced without any cell death (Duncan and Saldanha, 2011). These findings suggest that glial aromatase induction after traumatic brain injury is a neuroimmune response to injury (Figure 3.2).

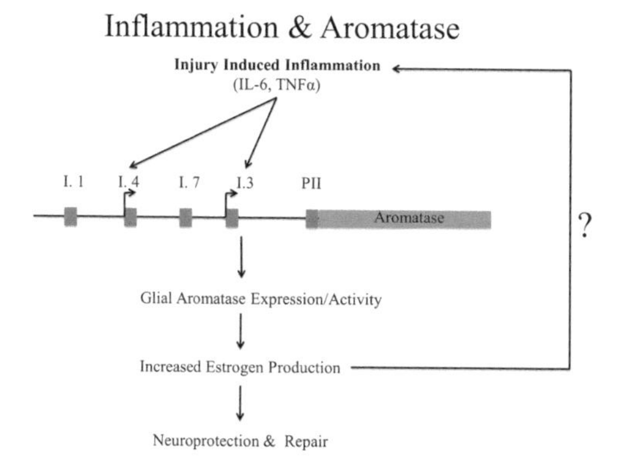

FIGURE 3.2 Current model of injury-induced aromatase production on neuroprotection in the avian brain. Briefly, following injury, aromatase is upregulated in glial cells around the injury site. This upregulation is hypothesized to be due to increases in proinflammatory cytokines that activate promoters on the aromatase gene within glial cells. The subsequent production of estrogens promotes neuroprotection via various mechanisms including increased neurogenesis, decreased apoptosis, and increased expression of developmental proteins. The culmination of these various processes is increased neuroprotection and neural and behavioral recovery, but the precise mechanism is still unknown (?).

Sex Differences in Cytokine Response Following Injury

As mentioned previously, it has been shown in other tissues that estrogen induction and the immune system are interrelated, in that estrogen levels can impact cytokine production, and cytokines can induce aromatization. Given the obvious differences in circulating hormone levels between males and females, and also the changes in hormone levels during menopause, it is not surprising that there are sex differences in the inflammatory response following injury. In fact, the neuroprotective effects of estrogen likely contribute to sex differences in neurodegenerative disease, ischemic stroke and traumatic brain injury. The neuroprotection in females is conferred in part by estrogen modulation of the immune response (Czlonkowska et al., 2006).

In some models of neurodegeneration, such as a mitochondrial toxin model of Parkinson's disease, the kinetics of cytokine expression differ between males and females (Czlonkowska et al., 2006). In astrocytes specifically, estrogen and selective estrogen receptor modulators influence cytokine levels (Cerciat et al., 2010; Dodel et al., 1999). While not many studies have looked at sex differences in the timeline of cytokine induction after injury, in songbirds females have a greater upregulation of IL-1β, but not IL-6, that occurs prior to a robust upregulation of aromatase that also has a female bias (Saldanha et al., 2013).

Given that males and females have different inflammatory responses to injury, and aromatase is regulated by inflammatory cytokines, these pathways may play an important role in sex differences in injury outcome.

Sex Differences in Glial Aromatase Expression Following Injury

Various groups have shown sex differences in the induction of glial aromatase following injury. In songbirds and in mammals, inhibition of aromatase using fadrozole increases damage after traumatic brain injury in males, an effect that can be prevented with estrogen replacement (Azcoitia et al., 2001; Wynne et al., 2008). In the zebra finch, there is a greater upregulation of aromatase mRNA expression that is more rapid in females than in males after damage to the cerebellum (Mirzatoni et al., 2010). A similar female-biased sex difference in glial aromatase induction was also observed after injury to the zebra finch entopallium (Saldanha et al., 2013). In this model, there are also differences in the kinetics of IL-1β expression, described above, with females having a more rapid and robust increase in expression before and during the aromatase expression peak after injury.

There are also sex differences in aromatase induction specifically in astrocytes. Astrocytes isolated from female neonatal rat cortex have greater aromatase activity and expression than astrocytes from males. Female astrocytes are protected from cell death in response to oxygen glucose deprivation compared to males, and this sex difference is abolished by aromatase inhibition (Liu et al., 2007).

Importantly, independently of brain injury, there are also sex differences in aromatase activity in different brain areas in zebra finches and in rats (Peterson et al., 2005; Roselli and Resko, 1993). The cause of these differences, though likely due to sex hormones, has not yet been fully elucidated.

CONCLUSION AND FUTURE DIRECTIONS

In recent years, the neuroprotective effects of estrogen have been elucidated in many models. The evidence described previously emphasizes the complex interactions of the neuroimmune and neuroendocrine systems in the induction of hormone synthesis after injury and the unique role for glial cells following injury. The finding that glial cells are the major players in steroid induction after injury has been enlightening, adding to the increasing amounts of evidence that glial cells are more than just the support for neurons. Additionally, local steroid hormone production by glia is part of a complex response to injury and can have profound physiological consequences. Given the increasing focus on and understanding of the sex differences and the role of menopause and hormone replacement therapies in brain injury outcome, it is not surprising that gender differences in this induction are present.

In the future, there remain several outstanding questions about this process. Further studies are required to address the role of the estrogen receptors in the induction and response to estrogen in traumatic brain injury. In the brain, estrogen receptors alpha, beta, and GPER are expressed in both neurons and glia and are thought to be involved in many different aspects of estrogen signaling via their nuclear and membrane associated localization (Hazell et al., 2009; Laflamme et al., 1998). Equally important is a full understanding of the neuroprotective signaling pathways that are triggered by local induction of estrogen after injury. In the next chapter, the role of glial aromatization in neuroprotection will be discussed in detail.

Acknowledgements

NIH NS 042767 supported this work. We thank Colin Saldanha for his mentorship and assistance in conducting several of the experiments described here. We also thank John M. Duncan for illustrations.

References

Acaz-Fonseca, E., Sanchez-Gonzalez, R., Azcoitia, I., Arevalo, M.A., Garcia-Segura, L.M., 2014. Role of astrocytes in the neuroprotective actions of 17beta-estradiol and selective estrogen receptor modulators. Mol. Cell. Endocrinol. 389 (1–2), 48–57.

Alvarez-Buylla, A., Kirn, J.R., 1997. Birth, migration, incorporation, and death of vocal control neurons in adult songbirds. J. Neurobiol. 33 (5), 585–601.

Arevalo, M.A., Santos-Galindo, M., Acaz-Fonseca, E., Azcoitia, I., Garcia-Segura, L.M., 2013. Gonadal hormones and the control of reactive gliosis. Horm. Behav. 63 (2), 216–221.

Azcoitia, I., Garcia-Ovejero, D., Chowen, J.A., Garcia-Segura, L.M., 2001. Astroglia play a key role in the neuroprotective actions of estrogen. Prog. Brain Res. 132, 469–478.

Bukulmez, O., Hardy, D.B., Carr, B.R., Word, R.A., Mendelson, C.R., 2008. Inflammatory status influences aromatase and steroid receptor expression in endometriosis. Endocrinology. 149 (3), 1190–1204.

Burda, J.E., Sofroniew, M.V., 2014. Reactive gliosis and the multicellular response to CNS damage and disease. Neuron. 81 (2), 229–248.

Cekanaviciute, E., Fathali, N., Doyle, K.P., Williams, A.M., Han, J., Buckwalter, M.S., 2014. Astrocytic transforming growth factor-beta signaling reduces subacute neuroinflammation after stroke in mice. Glia. 62 (8), 1227–1240.

Cerciat, M., Unkila, M., Garcia-Segura, L.M., Arevalo, M.A., 2010. Selective estrogen receptor modulators decrease the production of interleukin-6 and interferon-gamma-inducible protein-10 by astrocytes exposed to inflammatory challenge in vitro. Glia. 58 (1), 93–102.

Che, Q., Liu, B.Y., Liao, Y., Zhang, H.J., Yang, T.T., He, Y., et al., 2014. Activation of a positive feedback loop involving IL-6 and aromatase promotes intratumoral 17beta-estradiol biosynthesis in endometrial carcinoma microenvironment. Int. J. Cancer. 135 (2), 282–294.

Chen, Y., Swanson, R.A., 2003. Astrocytes and brain injury. J. Cereb. Blood Flow. Metab. 23 (2), 137–149.

Czlonkowska, A., Ciesielska, A., Gromadzka, G., Kurkowska-Jastrzebska, I., 2006. Gender differences in neurological disease: role of estrogens and cytokines. Endocrine. 29 (2), 243–256.

Dawson, V.L., Dawson, T.M., 1998. Nitric oxide in neurodegeneration. Prog. Brain Res. 118, 215–229.

DeFelipe, J., 2013. Cajal and the discovery of a new artistic world: the neuronal forest. Prog. Brain Res. 203, 201–220.

Dodel, R.C., Du, Y., Bales, K.R., Gao, F., Paul, S.M., 1999. Sodium salicylate and 17beta-estradiol attenuate nuclear transcription factor NF-kappaB translocation in cultured rat astroglial cultures following exposure to amyloid A beta(1-40) and lipopolysaccharides. J. Neurochem. 73 (4), 1453–1460.

Duncan, K.A., Saldanha, C.J., 2011. Neuroinflammation induces glial aromatase expression in the uninjured songbird brain. J. Neuroinflammation. 8, 81.

Elkabes, S., DiCicco-Bloom, E.M., Black, I.B., 1996. Brain microglia/macrophages express neurotrophins that selectively regulate microglial proliferation and function. J. Neurosci. 16 (8), 2508–2521.

Endoh, M., Maiese, K., Pulsinelli, W.A., Wagner, J.A., 1993. Reactive astrocytes express NADPH diaphorase in vivo after transient ischemia. Neurosci. Lett. 154 (1–2), 125–128.

Episcopo, F.L., Tirolo, C., Testa, N., Caniglia, S., Morale, M.C., Marchetti, B., 2013. Reactive astrocytes are key players in nigrostriatal dopaminergic neurorepair in the MPTP mouse model of Parkinson's disease: focus on endogenous neurorestoration. Curr. Aging Sci. 6 (1), 45–55.

Ferrero, S., Remorgida, V., Maganza, C., Venturini, P.L., Salvatore, S., Papaleo, E., et al., 2014. Aromatase and endometriosis: estrogens play a role. Ann. N. Y. Acad. Sci. 1317, 17–23.

Garcia-Segura, L.M., Chowen, J.A., Naftolin, F., 1996. Endocrine glia: roles of glial cells in the brain actions of steroid and thyroid hormones and in the regulation of hormone secretion. Front. Neuroendocrinol. 17 (2), 180–211.

Garcia-Segura, L.M., Naftolin, F., Hutchison, J.B., Azcoitia, I., Chowen, J.A., 1999a. Role of astroglia in estrogen regulation of synaptic plasticity and brain repair. J. Neurobiol. 40 (4), 574–584.

Garcia-Segura, L.M., Wozniak, A., Azcoitia, I., Rodriguez, J.R., Hutchison, R.E., Hutchison, J.B., 1999b. Aromatase expression by astrocytes after brain injury: implications for local estrogen formation in brain repair. Neuroscience. 89 (2), 567–578.

Ghirnikar, R.S., Lee, Y.L., Eng, L.F., 1998. Inflammation in traumatic brain injury: role of cytokines and chemokines. Neurochem. Res. 23 (3), 329–340.

Graeber, M.B., Streit, W.J., 2010. Microglia: biology and pathology. Acta. Neuropathol. 119 (1), 89–105.

Hazell, G.G., Yao, S.T., Roper, J.A., Prossnitz, E.R., O'Carroll, A.M., Lolait, S.J., 2009. Localisation of GPR30, a novel G protein-coupled oestrogen receptor, suggests multiple functions in rodent brain and peripheral tissues. J. Endocrinol. 202 (2), 223–236.

Hewett, S.J., Csernansky, C.A., Choi, D.W., 1994. Selective potentiation of NMDA-induced neuronal injury following induction of astrocytic iNOS. Neuron. 13 (2), 487–494.

Hill, R.A., Chua, H.K., Jones, M.E., Simpson, E.R., Boon, W.C., 2009. Estrogen deficiency results in apoptosis in the frontal cortex of adult female aromatase knockout mice. Mol. Cell. Neurosci. 41 (1), 1–7.

Honma, S., Shimodaira, K., Shimizu, Y., Tsuchiya, N., Saito, H., Yanaihara, T., et al., 2002. The influence of inflammatory cytokines on estrogen production and cell proliferation in human breast cancer cells. Endocr. J. 49 (3), 371–377.

Iadecola, C., Zhang, F., Casey, R., Nagayama, M., Ross, M.E., 1997. Delayed reduction of ischemic brain injury and neurological deficits in mice lacking the inducible nitric oxide synthase gene. J. Neurosci. 17 (23), 9157–9164.

Johann, S., Beyer, C., 2013. Neuroprotection by gonadal steroid hormones in acute brain damage requires cooperation with astroglia and microglia. J. Steroid Biochem. Mol. Biol. 137, 71–81.

Laflamme, N., Nappi, R.E., Drolet, G., Labrie, C., Rivest, S., 1998. Expression and neuropeptidergic characterization of estrogen receptors (ERalpha and ERbeta) throughout the rat brain: anatomical evidence of distinct roles of each subtype. J. Neurobiol. 36 (3), 357–378.

Lenzlinger, P.M., Morganti-Kossmann, M.C., Laurer, H.L., McIntosh, T.K., 2001. The duality of the inflammatory response to traumatic brain injury. Mol. Neurobiol. 24 (1–3), 169–181.

Liu, M., Hurn, P.D., Roselli, C.E., Alkayed, N.J., 2007. Role of P450 aromatase in sex-specific astrocytic cell death. J. Cereb. Blood Flow. Metab. 27 (1), 135–141.

Lull, M.E., Block, M.L., 2010. Microglial activation and chronic neurodegeneration. Neurotherapeutics. 7 (4), 354–365.

Malatesta, P., Gotz, M., 2013. Radial glia—from boring cables to stem cell stars. Development. 140 (3), 483–486.

Marciano, P.G., Eberwine, J.H., Ragupathi, R., Saatman, K.E., Meaney, D.F., McIntosh, T.K., 2002. Expression profiling following traumatic brain injury: a review. Neurochem. Res. 27 (10), 1147–1155.

Michels, M., Danielski, L.G., Dal-Pizzol, F., Petronilho, F., 2014. Neuroinflammation: microglial activation during sepsis. Curr. Neurovasc. Res.

Mirzatoni, A., Spence, R.D., Naranjo, K.C., Saldanha, C.J., Schlinger, B.A., 2010. Injury-induced regulation of steroidogenic gene expression in the cerebellum. J. Neurotrauma. 27 (10), 1875–1882.

Peterson, R.S., Fernando, G., Day, L., Allen, T.A., Chapleau, J.D., Menjivar, J., et al., 2007. Aromatase expression and cell proliferation following injury of the adult zebra finch hippocampus. Dev. Neurobiol. 67 (14), 1867–1878.

Peterson, R.S., Lee, D.W., Fernando, G., Schlinger, B.A., 2004. Radial glia express aromatase in the injured zebra finch brain. J. Comp. Neurol. 475 (2), 261–269.

Peterson, R.S., Saldanha, C.J., Schlinger, B.A., 2001. Rapid upregulation of aromatase mRNA and protein following neural injury in the zebra finch (*Taeniopygia guttata*). J. Neuroendocrinol. 13 (4), 317–323.

Peterson, R.S., Yarram, L., Schlinger, B.A., Saldanha, C.J., 2005. Aromatase is pre-synaptic and sexually dimorphic in the adult zebra finch brain. Proc. Biol. Sci. 272 (1576), 2089–2096.

Prinz, M., Mildner, A., 2011. Microglia in the CNS: immigrants from another world. Glia. 59 (2), 177–187.

Purohit, A., Ghilchik, M.W., Duncan, L., Wang, D.Y., Singh, A., Walker, M.M., et al., 1995. Aromatase activity and interleukin-6 production by normal and malignant breast tissues. J. Clin. Endocrinol. Metab. 80 (10), 3052–3058.

Purohit, A., Ghilchik, M.W., Leese, M.P., Potter, B.V., Reed, M.J., 2005. Regulation of aromatase activity by cytokines, PGE2 and 2-methoxyoestrone-3-O-sulphamate in fibroblasts derived from normal and malignant breast tissues. J. Steroid Biochem. Mol. Biol. 94 (1–3), 167–172.

Ramachandran, B., Schlinger, B.A., Arnold, A.P., Campagnoni, A.T., 1999. Zebra finch aromatase gene expression is regulated in the brain through an alternate promoter. Gene. 240 (1), 209–216.

Reed, M.J., Topping, L., Coldham, N.G., Purohit, A., Ghilchik, M.W., James, V.H., 1993. Control of aromatase activity in breast cancer cells: the role of cytokines and growth factors. J. Steroid Biochem. Mol. Biol. 44 (4–6), 589–596.

Robel, S., Berninger, B., Gotz, M., 2011. The stem cell potential of glia: lessons from reactive gliosis. Nat. Rev. Neurosci. 12 (2), 88–104.

Roselli, C.E., Resko, J.A., 1993. Aromatase activity in the rat brain: hormonal regulation and sex differences. J. Steroid Biochem. Mol. Biol. 44 (4–6), 499–508.

Saldanha, C.J., Burstein, S.R., Duncan, K.A., 2013. Induced synthesis of oestrogens by glia in the songbird brain. J. Neuroendocrinol. 25 (11), 1032–1038.

Saldanha, C.J., Duncan, K.A., Walters, B.J., 2009. Neuroprotective actions of brain aromatase. Front. Neuroendocrinol. 30 (2), 106–118.

Shibuya, S., Miyamoto, O., Auer, R.N., Itano, T., Mori, S., Norimatsu, H., 2002. Embryonic intermediate filament, nestin, expression following traumatic spinal cord injury in adult rats. Neuroscience. 114 (4), 905–916.

Sild, M., Ruthazer, E.S., 2011. Radial glia: progenitor, pathway, and partner. Neuroscientist. 17 (3), 288–302.

Singh, A., Purohit, A., Duncan, L.J., Mokbel, K., Ghilchik, M.W., Reed, M.J., 1997. Control of aromatase activity in breast tumours: the role of the immune system. J. Steroid Biochem. Mol. Biol. 61 (3–6), 185–192.

Smith, G.T., Brenowitz, E.A., Wingfield, J.C., 1997. Seasonal changes in the size of the avian song control nucleus HVC defined by multiple histological markers. J. Comp. Neurol. 381 (3), 253–261.

Snell, R.S., 2009. Clinical Neuroanatomy for Medical Students, seventh ed. Lippincott Williams & Wilkins, Baltimore.

Sofroniew, M.V., 2009. Molecular dissection of reactive astrogliosis and glial scar formation. Trends Neurosci. 32 (12), 638–647.

Sofroniew, M.V., Vinters, H.V., 2010. Astrocytes: biology and pathology. Acta. Neuropathol. 119 (1), 7–35.
Sohrabji, F., 2014. Estrogen-IGF-1 interactions in neuroprotection: ischemic stroke as a case study. Front. Neuroendocrinol.
Streit, W.J., Conde, J.R., Fendrick, S.E., Flanary, B.E., Mariani, C.L., 2005. Role of microglia in the central nervous system's immune response. Neurol. Res. 27 (7), 685–691.
Streit, W.J., Miller, K.R., Lopes, K.O., Njie, E., 2008. Microglial degeneration in the aging brain—bad news for neurons? Front. Biosci. 13, 3423–3438.
Tower, D.B., Young, O.M., 1973. The activities of butyrylcholinesterase and carbonic anhydrase, the rate of anaerobic glycolysis, and the question of a constant density of glial cells in cerebral cortices of various mammalian species from mouse to whale. J. Neurochem. 20 (2), 269–278.
Walters, B.J., Alexiades, N.G., Saldanha, C.J., 2011. Intracerebral estrogen provision increases cytogenesis and neurogenesis in the injured zebra finch brain. Dev. Neurobiol. 71 (2), 170–181.
Walters, B.J., Saldanha, C.J., 2008. Glial aromatization increases the expression of bone morphogenetic protein-2 in the injured zebra finch brain. J. Neurochem. 106 (1), 216–223.
Wu, D., Shibuya, S., Miyamoto, O., Itano, T., Yamamoto, T., 2005. Increase of NG2-positive cells associated with radial glia following traumatic spinal cord injury in adult rats. J. Neurocytol. 34 (6), 459–469.
Wynne, R.D., Maas, S., Saldanha, C.J., 2008a. Molecular characterization of the injury-induced aromatase transcript in the adult zebra finch brain. J. Neurochem. 105 (5), 1613–1624.
Wynne, R.D., Saldanha, C.J., 2004. Glial aromatization decreases neural injury in the zebra finch (*Taeniopygia guttata*): influence on apoptosis. J. Neuroendocrinol. 16 (8), 676–683.
Wynne, R.D., Walters, B.J., Bailey, D.J., Saldanha, C.J., 2008b. Inhibition of injury-induced glial aromatase reveals a wave of secondary degeneration in the songbird brain. Glia. 56 (1), 97–105.
Xing, L., Goswami, M., Trudeau, V.L., 2014. Radial glial cell: critical functions and new perspective as a steroid synthetic cell. Gen. Comp. Endocrinol. pii: S0016-6480(14)00083-5.
Yuan, Y.M., He, C., 2013. The glial scar in spinal cord injury and repair. Neurosci. Bull. 29 (4), 421–435.
Zhang, Y., Zhang, S., Lu, H., Zhang, L., Zhang, W., 2014. Genes encoding aromatases in teleosts: evolution and expression regulation. Gen. Comp. Endocrinol. pii: S0016-6480 (14)00180-4.
Zupanc, G.K., Zupanc, M.M., 2006. New neurons for the injured brain: mechanisms of neuronal regeneration in adult teleost fish. Regen. Med. 1 (2), 207–216.

CHAPTER

4

Astrocytic Aromatization and Injury

Colin J. Saldanha, Carissa J. Mehos, Alyssa L. Pedersen, and William A. Wiggins

Departments of Biology, Psychology and the Center for Behavioral Neuroscience, American University, Washington, DC, USA

Our understanding of how steroid hormones are provided to neural circuits has undergone considerable evolution. While the role of the ovary in synthesizing and providing estradiol (E_2) to the entire body via the circulatory system was known for some time, it was only in the early 1980s that we discovered that this steroid could be synthesized in the brains of both sexes. This was a discovery of staggering importance. We now appreciated that aromatase (*estrogen synthase*) could be expressed in the brain, but more importantly, in discrete brain areas where, presumably, circulating testosterone (T) could be aromatized to locally high levels of E_2. Now, not only was the brain a source and a target for E_2, but the functional consequences of this discrete E_2 synthesis were profound. Localized hypothalamic aromatization was key in the organization of masculine circuits during development in rodents, and in the activation of these circuits to express masculine copulatory and aggressive behaviors during adulthood in rodents and birds. Indeed, the metabolism of circulating androgens like T into distinct estrogenic metabolites like E_2 within constrained neural loci transformed and expanded our appreciation of hormonal signaling from endocrine pathways to paracrine and autocrine mechanisms powerful enough to modulate behavior. More specifically, in murine rodents and galliforme birds, the hypothalamic aromatization of circulating T was critical for consummatory masculine sexual behavior, evidenced by the fact that site-specific aromatization at this locus increased behaviors such as mounts, struts, intromissions, and

K.A. Duncan (Ed): Estrogen Effects on Traumatic Brain Injury.
DOI: http://dx.doi.org/10.1016/B978-0-12-801479-0.00004-8

other aspects of copulation. The specific forebrain areas and cell types responsible for this aromatization, however, remained a mystery.

The development of *in situ* labeling techniques was key in finding the answers to both these questions. Studies that targeted mRNA and the immunoreactive protein for aromatase successfully revealed that this enzyme was expressed in several areas of the brain and that this distribution differed across species. In mammals and galliformes, aromatase was largely constrained to diencephalic loci including the hypothalamic preoptic area. Additionally, in mammals, while other brain areas such as the amygdala and hippocampus did express aromatase, this expression appeared to decrease from early development to adulthood. In stark contrast, in songbirds, aromatase expression was much more widespread including, but not limited to, diencephalic and telencephalic areas such as the medial preoptic nucleus, ventromedial nucleus, nucleus taeniae (medial amygdala), hippocampus, bed nucleus of the stria terminalis, and ventral tegmental area. Perhaps the two most salient ideas that emerged from these studies were that aromatization seemed to be an important neurochemical component of nuclei within the "social-behavior network," and that this enzyme was expressed exclusively in neurons in brains of endotherms. As we were about to find out, aromatase expression was neither limited to constrained limbic brain areas, nor was it confined to neuronal cells.

AROMATASE IS EXPRESSED IN ASTROGLIA

The aromatase transcript and enzyme activity was first demonstrated in glial cells in primary-dissociated cultures of hatchling zebra finches (*Taenopygia guttata*). In these studies, aromatase enzyme activity was maintained or even increased with the age of the culture and despite the disappearance of neuronal cells. This suggested that aromatase was likely expressed in nonneuronal cells and, further, pointed to the possibility that some factor associated with the age of the culture may be regulating the expression of aromatase. Indeed, *in situ* hybridization studies revealed the presence of aromatase mRNA in astrocyte-like cells in the glial mat of these cultures. This finding was later replicated in cultures made from rodent brain, suggesting a conserved pattern of aromatase expression in nonneuronal cells of the vertebrate brain, at least during development. We were soon to learn, however, that these *in vitro* observations had revealed a fundamental property of the endotherm brain. Specifically, the observed expression of glial aromatase was likely due to perturbation of the neuropil, a necessary event in the preparation of primary dissociated brain cultures!

Aromatase detected in glial cells *in vitro* is now understood to reflect the induction of this enzyme in this specific cell type by some factor(s) associated with disruption and damage to the brain. Aromatase transcription and translation are upregulated in the rodent and passerine brain following various forms of insult including excitotoxicity, ischemia, anoxia, and mechanical perturbation. The first description of injury-induced aromatase in glial cells *in vivo* was presented following the peripheral administration of excitotoxic agents and has since been replicated in other rodents and birds with consistent results. The induction of aromatase occurs in reactive astrocytes that are localized either around the site of mechanical damage or in brain areas particularly vulnerable to excitotoxic injury.

While the induction of aromatase in response to damage is observed in mammals and birds, there are some notable differences. Specifically, in the songbird, aromatase induction appears to be more rapid and robust. Further, in songbirds, when the injury is located in proximity to the ventricular zone, aromatase is induced in another type of astroglial cell, specifically in radial glia. To the best of our knowledge, aromatase in radial glia has not been demonstrated in the mammalian brain. Thus, the data suggest that aromatase induction in the passerine brain may be more rapid, robust, sustained, and occur in a greater set of nonneuronal cells. We believe these characteristics render the songbird an excellent model for the study of injury-induced aromatase expression in glial cells, and will focus the rest of this chapter on work done in passerines. By exploiting the dramatic nature of glial aromatase induction in the songbird brain, we are now beginning to understand the precise nature of the signal(s) responsible for the transcription and translation of the aromatase gene in astroglia.

WHAT INDUCES ASTROCYTIC AROMATASE EXPRESSION?

Since mechanical brain damage results in the expression of aromatase in astrocytes near the site of injury, we reasoned that factors upregulated during brain damage may be good candidates as regulators of aromatase transcription. Among the many catastrophic and dramatic physiological responses to brain injury, the activation of the immune system, associated secretion of cytokines and chemokines, and consequent inflammation and edema are well understood. We considered, therefore, that it might be possible that cytokine expression may be associated with the upregulation of glial aromatase in the passerine brain. This hypothesis was tested by unilaterally exposing the neuropil to either the inflammagen phytohemagglutinin (PHA) or vehicle. In the hemisphere exposed to inflammagen, PHA increased expression of the cytokines interleukin 1β and 6 (IL-1β and IL-6) four hours post-treatment, suggesting PHA-dependent mobilization of the immune system. PHA also induced glial aromatase 24 hours following surgery measurable at both the transcript and protein levels. Importantly, the induction of glial aromatase was specific to the hemisphere treated with PHA, as vehicle-treated contralateral hemispheres were lower in aromatase transcript expression, and only contained the constitutive pool of neuronal aromatase detected with immunocytochemistry. Indeed, double-label immunocytochemistry confirmed that the induction of cytokine expression occurred in microglial cells, while that of aromatase expression occurred in astrocytes. These data suggest that inflammation is sufficient to induce aromatase in glia; however, critical experiments that test the necessity of inflammation to induce glial aromatase expression remain to be conducted.

GLIAL ESTROGEN SYNTHESIS AND NEUROPROTECTION

In contrast to the inductive signals discussed above, considerable work has focused upon the consequences of glial aromatization on the brain. More specifically, the influence of estrogens on cell turnover including apoptosis, cytogenesis and neurogenesis has received considerable attention. Here, we will discuss the influence of injury and glial aromatase on apoptosis and neurogenesis in songbirds while referencing some of the important work done in mammalian models.

Apoptosis and Secondary Degeneration

In the mammalian brain, including that of humans, damage resulting from a number of causes results in a characteristic wave of apoptotic secondary degeneration. This wave is most apparent 24–48 hours post insult and, importantly, is thought to underlie the most apparent and readily observable behavioral effects of the damage. We tested the effect of brain damage on the wave of apoptotic degeneration in the songbird by observing the effects of a single mechanical stab wound on the area of degeneration at various time-points between 2 hours to 6 weeks later. Surprisingly, and in contrast to mammals, the area of degeneration appeared constrained and invariant across the time points investigated. However, when the identical experiment was done in the presence of the aromatase inhibitor fadrozole, not only was the area of degeneration increased, but the temporal and spatial pattern of the spread of degeneration was very similar to that in the mammalian brain. Specifically, upon the inhibition of induced aromatase, the halo of degeneration increased during the early hours post injury, and peaked between 24 and 72 hours post damage. Thereafter, the area of damage decreased back to control levels. Further studies established that the damage caused by mechanical damage in the songbird was due to the activation of apoptotic pathways rather than necrosis. Specifically, the number of apoptotic nuclei around the site of damage was sufficient to explain the total number of dying cells, underscoring the specificity of the influence of glial aromatization on apoptotic pathways. More generally, however, these data suggest that the upregulation of aromatase in the passerine brain is robust enough to completely dampen the wave of secondary degeneration that is diagnostic of traumatic brain injury and other forms of insult in the brains of mammals, including humans. This striking *lack* of a wave of secondary degeneration in the zebra finch brain suggests a profound influence of glial aromatization in this species as opposed to mammals, and may be a reflection of the dramatic neuroplasticity within the passerine brain.

It has become clearer that the influence of glial aromatization on indices of degeneration is due, specifically, to high levels of E_2 production at the site of injury. We caused mechanical damage and administered fadrozole to both lobes of adult zebra finches; however, we replaced E_2 only in one lobe. Thus, while induced aromatase was inhibited around sites of damage in both hemispheres, one injection site received a dose of exogenous E_2. We found that E_2 replacement had a profound effect on the halo of degeneration. Specifically, the area of damage, the number of dying cells, and the number of apoptotic nuclei were all substantially lower in the hemisphere that received E_2 relative to the contralateral lobe. We have therefore concluded that the induction of glial aromatase at the site of damage results in local elevations in neural E_2, which is available to affect the molecular pathways responsible for apoptotic secondary degeneration. Taken together, these data point to a critical role for estrogens produced by reactive astrocytes around the site of damage. This influence is very likely due to the estrogenic modulation of apoptotic genes and pathways.

Injury-Induced Cyto- and Neurogenesis

The influence of locally produced estrogens extends beyond the regulation of cell death. Indeed, we have long appreciated the role of estrogens on multiple aspects of cell turnover including cytogenesis and neurogenesis. Since peripheral estrogens have documented effects on increased cell birth and proliferation, we decided to test the role of glial E_2 on injury-induced cyto- and neurogenesis. In these studies, and as described previously, adult zebra finches were subjected to bilateral injuries and inhibition of aromatase via the administration of fadrozole. However, E_2 was replaced into just one hemisphere. In two areas of the proliferative ventricular zone and around the injury, we counted the number of new cells and compared these numbers across hemispheres. There were minimal effects of E_2 replacement around the actual site of damage; however, in the subventricular zone, the lobe that received E_2 had significantly more proliferating cells. Importantly, the subset of these new cells that were neurons was also higher in the ventricular zone of the lobe that received E_2. The findings that estrogen signaling can contribute to neurogenesis in the subventricular zone following injury are tantalizing because they suggest that locally induced E_2 synthesis could influence neuroregeneration, and, importantly, local estrogen synthesis may be used to promote therapeutic regeneration by further augmenting neurogenesis.

SUMMARY AND FUTURE DIRECTIONS

Circulating estrogens have long been appreciated as neuroprotective agents in many animals, including humans. More recently, the discovery of injury-induced aromatase expression has focused our attention on the considerable effect of locally produced estrogens via glial aromatization on various indices of neuroprotection including cell proliferation and cell death. More specifically, several types of damage to the brains of birds and mammals result in the induction of aromatase transcription and translation in reactive astrocytes immediately around the site of damage. This induction appears to involve inflammatory pathways, as treatment with inflammatory agents is sufficient to increase microglial cytokines and astrocytic aromatase. Microglial secretions including cytokines and trophic factors may be important mediators of the neuroprotective effects of astroglial aromatization. There is some suggestion that the dynamics of aromatase induction and the consequences of glial estrogen provision may be more dramatic in the songbird compared to the mammal; however, there is considerable agreement as to the influence of glial estrogens on pathways that are neuroprotective in birds and mammals.

Further Reading

Azcoitia, I., Sierra, A., Veiga, S., Garcia-Segura, L.M., 2003. Aromatase expression by reactive astroglia is neuroprotective. Ann. N.Y. Acad. Sci. 1007, 298–305.

Duncan, K.A., Walters, B.J., Saldanha, C.J., 2013. Traumatized and inflamed, but resilient: glial aromatization and the avian brain. Horm Behav. 83 (2), 208–215.

Garcia-Segura, L.M., McCarthy, M.M., 2004. Minireview: role of glia in neuroendocrine function. Endocrinology. 145 (3), 1082–1086.

Garcia-Segura, L.M., Veiga, S., Sierra, A., Melcangi, R.C., Azcoitia, I., 2003. Aromatase: a neuroprotective enzyme. Prog. Neurobio. 71 (1), 31–41.

Saldanha, C.J., Burstein, S.R., Duncan, K.A., 2013. Induced synthesis of oestrogens by glia in the songbird brain. J. Neuroendocrinol. 25 (11), 1032–1038.

Saldanha, C.J., Duncan, K.A., Walters, B.J., 2009. Neuroprotective actions of brain aromatase. Front Neuroendocrinol. 30 (2), 106–118.

CHAPTER

5

Aromatase and Estrogens: Involvement in Constitutive and Regenerative Neurogenesis in Adult Zebrafish

Elisabeth Pellegrini[1], *Pascal Coumailleau*[1], *Olivier Kah*[1], *and Nicolas Diotel*[2]

[1]Neuroendocrine Effects of Endocrine Disruptors, IRSET, INSERM U1085 Université de Rennes, Rennes, France [2]INSERM U1188, Plateforme CYROI, Université de La Réunion, Saint Denis de La Réunion, France

INTRODUCTION

Nowadays, with the increase of brain strokes and neurodegenerative diseases, a better understanding of mechanisms triggering constitutive and regenerative neurogenesis is a key challenge for finding treatments and cures. For a long time, estrogens have been known to exert pleiotropic effects on a wide range of tissues and organs. In the central nervous system (CNS), they have neuroprotective properties and modulate neurogenesis, brain repair and synaptic plasticity. In this context, teleost fish exhibit interesting and almost unique features in vertebrates. Among teleosts, the zebrafish (*Danio rerio*) is a useful tool for drug screening and a recognized model organism for studying a wide variety of physiological processes and disorders such as immunity, cancer, neurogenesis and regeneration (Santoriello and Zon, 2012). In this chapter, we aim to emphasize the contribution of fish, and peculiarly zebrafish, in the better understanding of the roles of aromatase and estrogens in

K.A. Duncan (Ed): Estrogen Effects on Traumatic Brain Injury.
DOI: http://dx.doi.org/10.1016/B978-0-12-801479-0.00005-X

constitutive and reparative neurogenesis. First of all, we will document the unique features exhibited by the brain of teleost fish in terms of aromatase activity, neurogenesis and regenerative capability. We will next focus on neurosteroid synthesis and signaling in the brain of adult zebrafish, and discuss the fact that radial glial cells (RGCs) could be a source of steroids and a target of peripheral and locally produced steroids. Then, we will highlight the role of estrogens in constitutive neurogenesis and their potential involvement in brain repair. Finally, we will compare these data with those of other vertebrates, with a special focus on amphibians.

UNIQUE FEATURES EXHIBITED BY THE BRAIN OF TELEOST FISH

Among vertebrates, teleost fish are unique in many respects and emerge as fascinating models especially in the field of neurosciences (Figure 5.1).

Firstly, pioneer studies in the 1980s documented the presence of high aromatase activity in the brain of adult goldfish and toadfish, reported to be 100 to 1000 times higher than in the corresponding regions in

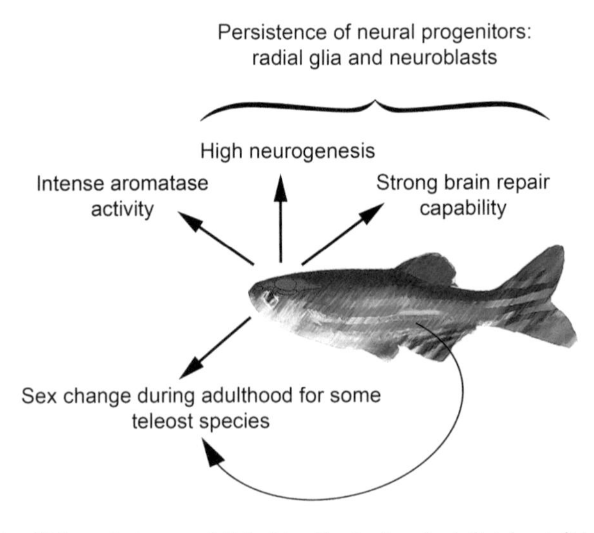

FIGURE 5.1 Unique features exhibited by the brain of adult teleost fish. The brain of adult teleost fish displays a strong aromatase activity and a high constitutive and regenerative neurogenesis allowed by the persistence of neural progenitors in the brain (RGCs and further committed precursors). Some teleost species have the capacity to change sex during the lifespan, implying a strong gonadal and neuronal plasticity.

mammals (Callard et al., 1990; Pasmanik and Callard, 1985). This strong aromatase activity was next confirmed in many other species including zebrafish, and was shown to result from the intense expression of a brain-specific aromatase, aromatase B (AroB), encoded by the *cyp19a1b* gene, one of the two *cyp19a1* genes emerging from a genomic duplication event occurring some 325 million years ago in teleost fish (Diotel et al., 2010a; Le Page et al., 2010; Ravi and Venkatesh, 2008). In contrast to other vertebrates in which brain aromatase is mainly expressed in neurons, it is only expressed in radial glial cells (RGCs) in the brain of fish (see section titled "Aromatase B Expression Is Restricted to Radial Glia Acting Like Neural Stem Cells").

Secondly, the brain of teleost fish also displays a widespread and intense neurogenesis during the lifespan (Zupanc, 2001). While there are only two main neurogenic regions in mammals, the dentate gyrus of the hippocampus and the subventricular zone (SVZ) of the lateral ventricle, the brain of adult teleosts exhibits a high density of neurogenic niches throughout the whole brain (Lindsey and Tropepe, 2006; Pellegrini et al., 2007). This strong neurogenic activity is due to the persistence and activity of RGCs during adulthood as well as further committed progenitors (März et al., 2010; Onteniente et al., 1983; Pellegrini et al., 2007). The fact that RGCs are the only ones expressing the estrogen-synthesizing enzyme highlights substantially the potential involvement for aromatase and estrogens in zebrafish neurogenesis (see sections "Aromatase B Expression Is Restricted to Radial Glia Acting Like Neural Stem Cells" and "Involvement of Aromatase and Estrogens in Adult Neurogenesis and Brain Repair in Zebrafish").

Furthermore, in contrast to mammals, teleost fishes exhibit a remarkable ability to regenerate nervous tissue by a process involving extensive neurogenic events (Grandel and Brand, 2013; Kizil et al., 2012). In the last few years, this injury-induced neurogenesis has been intensively studied in zebrafish (Becker and Becker, 2008; Schmidt et al., 2013), and estrogens were suggested to be key players in brain remodeling and regenerative processes (Diotel et al., 2013) (see section "Involvement of Aromatase and Estrogens in Adult Neurogenesis and Brain Repair in Zebrafish").

Last but not least, some teleost fishes show an intriguing capacity to change sex during life (Le Page et al., 2010). This implies that gonad determination and differentiation differ in fish compared to other vertebrates (Guiguen et al., 2010; Nagahama, 2005). It also means that neuronal networks controlling gonadotrophin release, reproductive status and behavior are plastic and not permanently sexualized in contrast to mammals, and that aromatase and estrogens could be key players in such processes (Le Page et al., 2010).

NEUROSTEROID SYNTHESIS AND SIGNALING IN THE BRAIN OF ADULT ZEBRAFISH

Until recently, neurosteroid synthesis was poorly documented in teleost fish (Diotel et al., 2011a,b). Given the high aromatase activity exhibited by the brain of adult zebrafish, the question of local versus peripheral origin of C19 androgens available for aromatization was raised. In 2011, it was shown that the brain of adult male and female zebrafish was able to convert [^{3}H]-pregnenolone into a wide variety of radiolabeled steroids, namely testosterone, progesterone, estrone and estradiol (Diotel et al., 2011a). This study also demonstrates that the main steroidogenic enzymes leading to estrogen synthesis (*cyp11a1* -P450$_{SCC}$-, *3β-hsd*, *cyp17* and *cyp19a1b*) are biologically active and display a similar distribution in numerous brain regions. Interestingly, *cyp11a1*, *3β-hsd* and *cyp17* mRNAs are expressed in some AroB$^+$ RGCs. Apart from aromatase, the neuronal expression of these enzymes is not excluded given their wide distribution. This is further reinforced by the fact that 3βHSD-like immunoreactivity was observed in neurons throughout the adult zebrafish brain (Sakamoto et al., 2001). While these data allow us to consider the brain of adult zebrafish as a true steroidogenic organ (Figure 5.2), the relative and respective abundance of peripheral versus locally produced steroids that impregnate the brain are still unknown. It also raises the question of the targets of steroids in the brain. We will consequently discuss the current knowledge of the expression of nuclear and membrane associated steroid receptors for androgens, progestins and estrogens in the brain of adult zebrafish.

In zebrafish, androgen receptor (AR) was documented to actively bind testosterone (T), 5α-dihydro-T, 11-keto-T, and androstenedione (Jorgensen et al., 2007). In the adult brain, *ar* transcripts were notably detected in the parenchyma but also along the ventricular layers of the preoptic area, hypothalamus and of the periglomerular gray zone of the optic tectum (Gorelick et al., 2008). This distribution strongly argues in favor of AR expression in both neurons and RGCs (Figure 5.2), implying that androgens may possibly modulate the activity of RGCs.

Concerning progestagen signaling, a unique progesterone receptor (Pgr) was characterized in zebrafish. It actively binds progesterone (P), 17-hydroxy-P, dihydro-P, and 4-pregnen-17,20β-diol-3-one (Chen et al., 2010; Hanna et al., 2010). In zebrafish brain, Pgr expression was described at the mRNA and protein levels in both neurons and RGCs. These latest display a stronger Pgr expression than neuronal cells (Figure 5.2), suggesting that RGCs could be preferential targets for progesterone (Diotel et al., 2011c). This is of particular interest given the accumulating studies in mammals showing a role of progesterone in

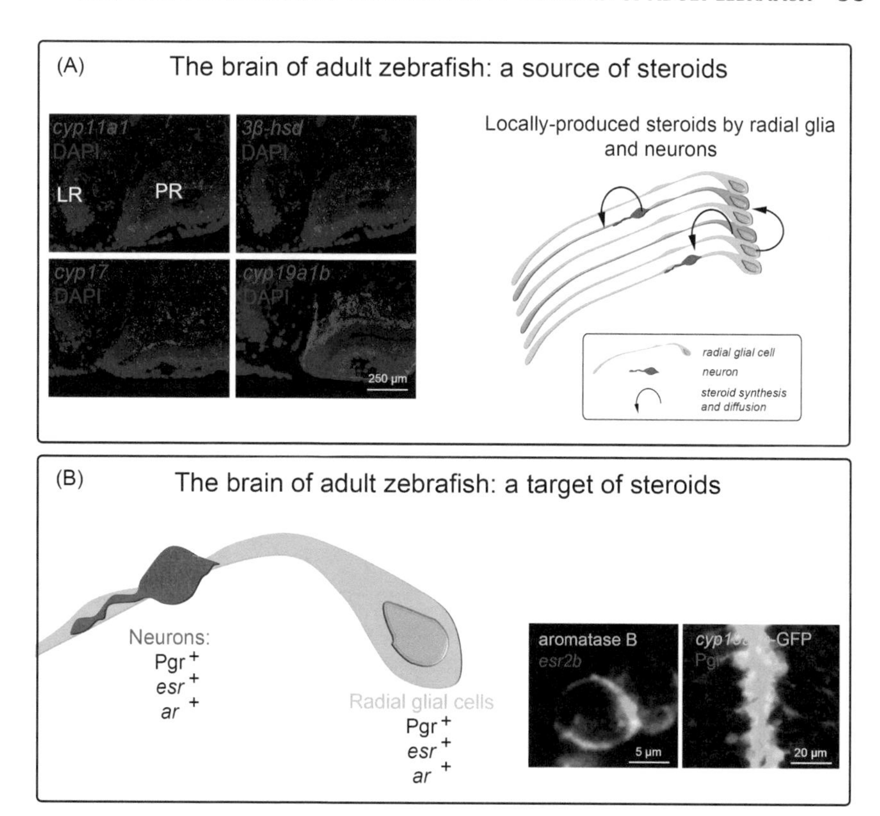

FIGURE 5.2 The brain of fish is a source and a target of steroids. A: *In situ hybridization* for *cyp11a1* ($P450_{scc}$), *cyp17*, *3β-hsd* and *cyp19a1b* (red) shows a similar distribution around the lateral (LR) and posterior recess (PR) of the hypothalamus. Cell nuclei are counterstained with DAPI (blue). The scheme illustrates neurosteroid synthesis by neurons and RGCs (arrows). B: Accumulating data show that neurons express classical nuclear receptor for progesterone (Pgr), estrogens (*esr1*, *esr2a* and *esr2b*) and androgens (*ar*). RGCs also express Pgr and probably *ar*, *esr2a* and *esr2b*. A weak *esr2b* mRNA expression is notably detected in $AroB^+$ RGCs and a stronger Pgr expression in *cyp19a1b*-GFP expressing RGCs such as shown in the hypothalamus for both pictures.

neurogenesis (Barha et al., 2011; Zhang et al., 2010). Interestingly, a recent transcriptomic analysis also showed that *pgr* is an estrogen target gene in zebrafish (Hao et al., 2013). It explains that 17β-estradiol upregulates Pgr expression in the brain of larvae and adult zebrafish, by increasing notably the number of Pgr-positive RGCs and Pgr relative expression (Diotel et al., 2011c). Consistently, inhibition of estrogen synthesis leads to a significant decrease of *pgr* expression in the adult brain. Such data reinforce the idea that estrogens could regulate progesterone-mediated effects on RGCs by indirectly modulating Pgr expression.

New data about membrane progestin receptors (mPRα, β and γ) expression recently emerged in zebrafish (Hanna and Zhu, 2009). However, their expression and roles are poorly documented, notably in the CNS in which only pituitary expression was described (Hanna and Zhu, 2009).

Finally, the three zebrafish nuclear estrogen receptor transcripts *esr1* (ERα), *esr2a* (ERβ2) and *esr2b* (ERβ1) were shown to be widely expressed in the brain of adult zebrafish (Diotel et al., 2011b; Menuet et al., 2002; Pellegrini et al., 2005). *In situ* hybridization (ISH) for *esr1, esr2a* and *esr2b* show their expression in the subpallium, the anterior and posterior part of the preoptic area and also in the whole hypothalamus. They are detected in neurons and along the ventricular layer, suggesting their possible expression in RGCs. Preliminary data notably reinforce this point showing *esr2b* expression in $AroB^+$ RGCs (Figure 5.2). It would be consistent with *in vivo* and *in vitro* results showing that aromatase expression is regulated by ERs through an estrogen responsive element (ERE) located at −348 bp of the *cyp19a1b* promoter (Menuet et al., 2005; Mouriec et al., 2009). In addition to the classical signaling through nuclear receptors, estrogens could also exert their effect by binding a membrane estrogen receptor called Gper or GPR30, corresponding to a G-protein-coupled receptor. In 2009, a wide *gper* expression was described in parenchymal, periventricular and ventricular cells in the brain of adult zebrafish (Liu et al., 2009), suggesting *gper* expression in RGCs. In addition, taking advantage of the *cyp19a1b*-GFP transgenic zebrafish line, it was shown that acute brain slice exposure to 17β-estradiol (10^{-7}M) results in a rapid modulation of RGCs and parenchymal cell activities, as revealed by calcium imaging (Pellegrini et al., 2013). So, estrogens could locally exert a wide range of effects modulating RGC activities and brain plasticity.

AROMATASE B EXPRESSION IS RESTRICTED TO RADIAL GLIA ACTING LIKE NEURAL STEM CELLS

Studies carried out over the last 20 years have managed to identify the nature of $AroB^+$ cells in the brains of various fish species. Using antibodies raised against human aromatase, pioneer investigations identified few positive neuronal cells in the most anterior part of the goldfish brain (Gelinas and Callard, 1997). However, the low number of labeled cells and their restricted distribution contrasted with the massive aromatase activity measured throughout the goldfish brain. The development of homologous molecular tools such as labeled riboprobes and specific antibodies subsequently opened the door to the investigation of the cerebral localization of aromatase-expressing cells in fish. It finally led to convincing documentation of the distribution of numerous AroB$^+$ cells along the ventricular layer in the forebrain and hindbrain of the midshipman first (Forlano et al., 2001), and next of many other teleost fishes such as rainbow trout (Menuet et al., 2003), zebrafish (Goto-Kazeto et al., 2004; Menuet et al., 2005; Pellegrini et al., 2005), pejerrey (Strobl-Mazzulla et al., 2005), killifish (Greytak et al., 2005), tilapia (Chang et al., 2005), and also medaka (Okubo et al., 2011). Recently, it was shown in the Japanese eel, a basal teleost fish displaying a single *cyp19a1* gene, that aromatase distribution was similar to that highlighted in teleost species possessing duplicated *cyp19a1* genes (Jeng et al., 2012). In zebrafish, AroB protein and mRNA overlap along the ventricles of the hindbrain, midbrain, diencephalon, telencephalon and olfactory bulbs (Menuet et al., 2005; Pellegrini et al., 2005).

Whereas neurons are the primary source of aromatase in the brain of mammals and birds (Balthazart and Ball, 1998; Peterson et al., 2005), AroB expression is restricted to RGCs and never expressed in neurons in the adult brain of teleost species studied to date (Diotel et al., 2010a). The radial glia nature of AroB-expressing cells was easily determined considering RGCs specificities including (1) a soma close to the ventricles and (2) a peculiar cell morphology characterized by a small cell body with an ovoid nucleus, a short process directed towards the ventricle and a long distal extension crossing the cerebral parenchyma and contacting the basal surface of the brain with end feet (Bentivoglio and Mazzarello, 1999; Rakic, 1978). In addition, convincing double-labeling experiments with AroB antibodies and canonical RGC markers such as GFAP, BLBP, GS and S100β definitively confirmed the radial glial identity of AroB-expressing cells in the brain of fish (Forlano et al., 2001; März et al., 2010; Menuet et al., 2003; Pellegrini et al., 2007; Tong et al., 2009). It was also reinforced by the fact that AroB expression never colocalized with Hu and acetylated tubulin neuronal markers (Forlano et al., 2001; Pellegrini et al., 2007). Curiously, in zebrafish, a strong *cyp19a1b* expression is also detected far away from the ventricular layer in radial processes, corresponding probably to *cyp19a1b* mRNA export in RGC processes far from the cell bodies (Diotel et al., 2011c; Menuet et al., 2005; Pellegrini et al., 2005).

In mammals and birds, RGCs were first described as a guide for newborn neurons migrating to their final destination (Rakic, 1990). However, their roles have been enlarged and their neural progenitor properties are now clearly established (Götz and Huttner, 2005). In mammals, at the end of brain development, most RGCs differentiate into astrocytes, but some are maintained in the neurogenic niches during adulthood (Brunne et al., 2010; Liu et al., 2006). However, in teleost fishes, RGCs persist during the lifespan and keep neurogenic capacities as we will describe (Onteniente et al., 1983; Pellegrini et al., 2007;

Zupanc and Clint, 2003). Among these RGCs, a huge majority express AroB (Forlano et al., 2001; Menuet et al., 2003; Nagarajan et al., 2011; Pellegrini et al., 2007). The specific ventricular position of $AroB^+$ RGCs highlighted in the previously mentioned works reminds one of the ventricular proliferation that has been documented in different fish species by using incorporation of bromodeoxyuridine (BrdU) or proliferating cell ($PCNA^+$) stainings (Adolf et al., 2006; Ekström et al., 2001; Grandel et al., 2006; Pellegrini et al., 2007; Strobl-Mazzulla et al., 2010; Zupanc and Horschke, 1995). In zebrafish, many ventricular neurogenic niches have been described to generate newborn cells that migrate away from the ventricle and differentiate in mature neurons (Adolf et al., 2006; Chapouton et al., 2007; Grandel et al., 2006). Our group demonstrated for the first time in zebrafish that a large subset of dividing cells corresponds to $AroB^+$ RGCs from the rostral to caudal part of the brain (Pellegrini et al., 2007). The newborn cells slowly migrate to their final destination by moving on the cytoplasmic extensions of $AroB^+$ RGCs and differentiate into mature neurons as evidenced by Hu and acetylated-tubulin stainings (Pellegrini et al., 2007). It was recently shown that in the paraventricular organ of the hypothalamus, a specific area of nonmammalian species, $AroB^+$ RGCs locally give birth to serotoninergic cerebrospinal fluid-contacting neurons (Pérez et al., 2013).

INVOLVEMENT OF AROMATASE AND ESTROGENS IN ADULT NEUROGENESIS AND BRAIN REPAIR IN ZEBRAFISH

The synthesis of estrogens by RGCs (Pellegrini et al., 2007) together with the well-known effects of estradiol in mammalian brain plasticity (Brock et al., 2010; Ormerod et al., 2004) raised the question of a functional relation between these hormones and cell fate in fish neurogenesis. Taking advantage of the zebrafish model, it was shown that Gper knock-down using morpholino results in an impaired neurogenesis during development, showing the importance of estrogen signaling in brain morphogenesis (Shi et al., 2013). In adult zebrafish, the potential role of estradiol in neurogenesis was recently investigated by manipulating the endogenous concentrations of the hormone (Diotel et al., 2013). When aromatase activity is blocked with ATD (10^{-6} M, 15 days), the number of proliferating cells ($PCNA^+$) increases throughout the forebrain, particularly in the olfactory bulbs/telencephalon junction, telencephalon, preoptic area, thalamus and mediobasal hypothalamus, even if this increase does not reach statistical significance. To further clarify the potential role of estradiol in proliferation, ERs were also blocked with $ICI_{182,780}$ (10^{-7} M, 54 h), resulting in a significant increase in proliferation at the olfactory bulbs/telencephalon junction and in the mediobasal part of the hypothalamus. Consistently, adult zebrafish treated with 17β-estradiol (10^{-7} M, 100 h) exhibit a significant decrease in the number of PCNA^{+} cells in the same brain regions (Diotel et al., 2013). Furthermore, 17β-estradiol was shown to negatively impact cell migration at the olfactory bulbs/telencephalon junction and in the mediobasal hypothalamus, as newborn cells accumulate close to the ventricle and decreases in the parenchyma. However, no clear data were obtained for cell survival and differentiation. Taken together, these data strongly suggest that 17β-estradiol negatively regulates adult neurogenesis in zebrafish in these experimental conditions (Figure 5.3), in contrast to what is generally described in mammals. New data obtained in zebrafish also highlighted a sexual dimorphism in brain cell proliferation in the dorsal telencephalon, the periventricular nucleus of the posterior tuberculum and the ventral pretectal area, mature females exhibiting a stronger number of cycling cells than males (Ampatzis et al., 2012). In contrast, in the dorsal part of the periventricular hypothalamus, the situation is the opposite, with a higher number of dividing cells in males compared to females (Ampatzis et al., 2012). These differences argue in

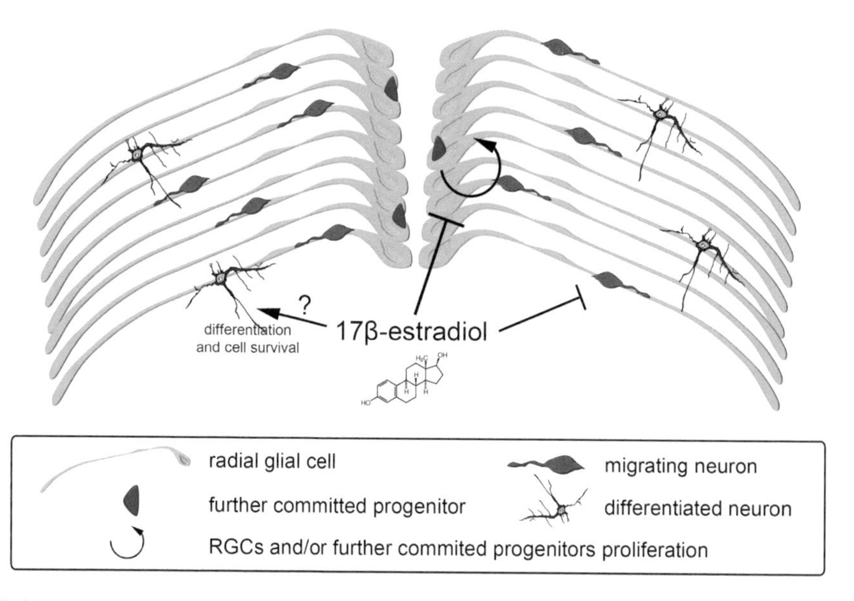

FIGURE 5.3 17β-estradiol modulates adult neurogenesis in zebrafish. Adult zebrafish exposure to 17β-estradiol (10^{-7} M for 100 h) leads to a significant decrease of brain cell proliferation in the ventricular region of the olfactory bulb/telencephalon junction and in the mediobasal hypothalamus. It also impairs neuronal migration after 14 and 28 days of treatment in the same regions. The effects on differentiation and cell survival are not so clear and would require further investigation.

favor of a key contribution of sexual steroids in neurogenic mechanisms. However, for AroB expression, no sexual dimorphism could be detected until now in the adult zebrafish brain, and further investigations would be required for the other stereoidogenic enzymes. In medaka, another fish model used in neurosciences, AroB expression is higher in females, especially in the optic tectum (Okubo et al., 2011). In females, the tectum express a higher level of genes involved in cell proliferation and a lower level of anti-apoptotic genes compared to males (Takeuchi and Okubo, 2013). Such data could imply that a differential level of estradiol resulting from AroB expression dimorphism could affect the cell life cycle in medaka.

In contrast to the compromised capacities of mammals to repair nervous structures after injury, nonmammalian vertebrates, and particularly teleost fish, possess a great power to regenerate damaged tissue by stimulating reparative neurogenesis mechanisms leading to almost complete recovery of the lost function (Schmidt et al., 2013; Zupanc, 2009). The zebrafish has recently emerged as a suitable model to address the question of cellular and molecular processes induced during the regeneration (Grandel and Brand, 2013; Kishimoto et al., 2012; März et al., 2011). As adult zebrafish regenerate neural tissue after injuries so efficiently, a model of lesions of the telencephalon has been developed in order to investigate the regenerative mechanisms and highlight the potential role of estradiol in this response (Diotel et al., 2013). As described in recent papers dedicated to reparative processes in zebrafish, a strong increase in proliferation is observed in the telencephalon after the lesion compared to the control side (Kishimoto et al., 2012; Kroehne et al., 2011; März et al., 2011). The surge in the number of proliferating cells noticeable from 24 hours in the injured parenchyma corresponds to oligodendrocytes and microglia recruitment at the injury site (März et al., 2011). At later times (from day 2 to 7), the number of dividing cells decreased in the parenchyma while more and more cycling cells appeared in the ventricular zones of the telencephalon. At these stages, the identity of the increasing number of ventricular proliferative cells in the injured telencephalon has been investigated using different markers such as AroB/BLBP/GFAP/S100β (RGC markers), Sox2 (neural progenitor marker) and PSA-NCAM (further committed progenitors marker). It allows characterizing the ventricular dividing cells as BLBP/GFAP/S100β$^+$ RGCs and as further committed progenitors, but not as AroB$^+$ RGCs (Baumgart et al., 2012; Diotel et al., 2013; März et al., 2011). Interestingly, at the same stages, the *cyp19a1b* gene and AroB expressions decrease in the injured telencephalon compared to the control hemisphere. This AroB downregulation correlates with the surge in proliferative activity and is in agreement with the inhibitory effect of estradiol on proliferation in homeostatic conditions. So, in order to evaluate the involvement of estradiol in the cellular events evidenced after mechanical lesion in the telencephalon, zebrafish have been treated with 17β-estradiol (10^{-7} M, for 48 h and 7 days) or with ICI. The results do not point to a significant impact on injury-induced proliferation in such experimental conditions (Diotel et al., 2013). However, the apparent absence of estradiol effect on proliferation after injury does not necessarily mean that survival or differentiation of newborn neurons generated after the lesion are not affected. Very interestingly, some $AroB^+$ cells were observed in brain parenchyma 72h after the telencephalon injury. The identity of these *de novo* AroB expressing cells is currently unknown and could correspond to neurons or oligodendrocytes (Diotel et al., 2013). Even if this point would require further investigation, it is quite relevant given that AroB is only detected in RGCs and never observed in the parenchymal cells in normal conditions.

COMPARISON WITH MAMMALIAN AND NONMAMMALIAN MODELS

In contrast to teleosts, mammals and birds exhibit only restricted to moderate constitutive and regenerative neurogenic capabilities, as well as a relatively weak aromatase activity during adulthood. In general, estrogens have been described for increasing neurogenesis by improving neural progenitor cell proliferation, differentiation and cell survival (Bowers et al., 2010; Brinton, 2009; Martinez-Cerdeño et al., 2006; Ormerod et al., 2004; Wang et al., 2003; Wang et al., 2008). Such data could appear in contrast to what was shown in adult zebrafish. However, recent works in mice show that estradiol negatively regulates neural progenitor proliferation in the SVZ (Brock et al., 2010; Veyrac and Bakker, 2011), and estrogens were shown to exert opposite effects on neuroplasticity and cognition according to their nature and concentration (Barha and Galea, 2010). Furthermore, under physiological conditions, aromatase is mainly expressed by neurons in mammals and birds. However, after chemical or mechanical lesions, aromatase is *de novo* expressed in reactive astrocytes in rats and birds (Azcoitia et al., 2003; Garcia-Segura et al., 1999; Peterson et al., 2001) but also in RGCs facing the lesion in birds (Peterson et al., 2004). This is in striking contrast to teleosts and notably zebrafish in which brain aromatase is only expressed in RGCs under physiological conditions, while a decreased expression in RGCs and a *de novo* expression is observed in parenchymal cells after mechanical injury of the telencephalon (Diotel et al., 2013). So far, it appears that similarities and discrepancies occur between models (Figure 5.4), and push researchers to better take into

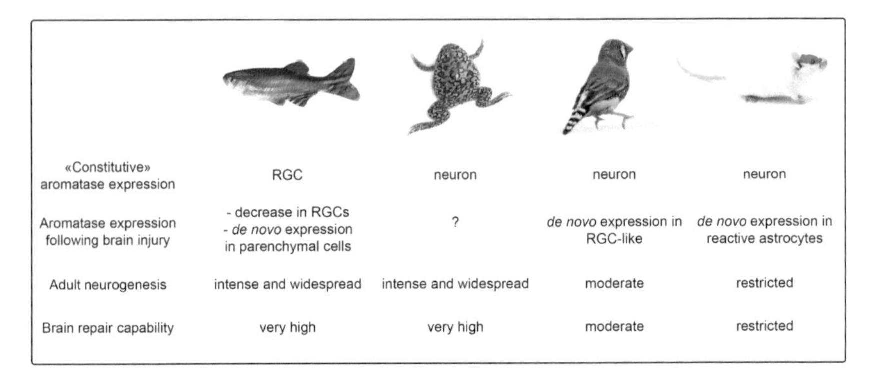

FIGURE 5.4 Similarities and discrepancies concerning brain aromatase expression and adult neurogenesis in homeostatic or injured conditions in teleost fish, amphibians, birds and rodents during adulthood.

consideration sex, estrogenic concentration and duration of treatment, brain structures analyzed and developmental stages studied for their experiments.

Among vertebrates, amphibians share some common features with teleost fish. They notably emerged as a leading model for studying constitutive and regenerative neurogenesis. Firstly, the brain of adult frog exhibits a widespread proliferation such as shown recently in *Xenopus laevis* (D'Amico et al., 2011), and gives rise to newborn neurons and oligodendrocytes (D'Amico et al., 2011; Simmons et al., 2008). Secondly, RGCs are also maintained during adulthood in many frog brain subdivisions (D'Amico et al., 2011, 2013). Finally, amphibians are also highly regenerative animals and have been used in various CNS regeneration studies (Tanaka and Ferretti, 2009). As well, the brain of adult amphibian produces neurosteroids, which might have important developmental or physiological roles in the CNS and in neurogenesis (Do Rego et al., 2009; Vaudry et al., 2011). The estrogen-synthesis enzyme P450 aromatase, encoded by one single *cyp19a1* gene, has been shown to exhibit a strong biological activity in the preoptic area and the hypothalamus of different adult frog species (Callard et al., 1978; Mensah-Nyagan et al., 1999), and aromatase-expressing cells were abundantly detected in these neuroendocrine regions in adult lizard (Cohen and Wade, 2011).

Interestingly, similar to what was described in fish, estrogens probably exert a positive autoregulatory feedback loop on aromatase expression given (1) the presence of a half ERE on aromatase promoter in frog and (2) that exposure to estrogens or ethinylestradiol increases *cyp19a1* expression in the brain (Akatsuka et al., 2005). Recently, ISH experiments precisely described the distribution of *cyp19a1* transcripts in the brain of *Xenopus laevis* during development and postmetamorphosis

(Coumailleau and Kah, 2014). From larval stage 42 until postmetamorphic stages, a very strong *cyp19a1* expression was detected in the preoptic area and the caudal hypothalamus (Coumailleau and Kah, 2014). While recently published data using mammalian aromatase antibodies further describes aromatase protein distribution in the frog brain (Burrone et al., 2012), its exact neuroanatomical distribution remains to be investigated using specific antibodies. However, in striking contrast to fish, under normal conditions, *cyp19a1* transcripts were exclusively detected in neurons, and not in RGCs or progenitor cells (Coumailleau and Kah, 2014). In *Xenopus Laevis* and *Xenopus tropicalis*, two (ERα, ERβ) or three (ERα1, ERα2, ERβ) isoforms of estrogen receptor have been identified respectively (Iwabuchi et al., 2008; Takase and Iguchi, 2007). In *Xenopus laevis* and in the urodele *Taricha granulosa*, a widespread distribution of estrogen-concentrating nerve cells was notably observed in the hypothalamic and limbic structures (Davis and Moore, 1996; Morrell et al., 1975). In addition, in the male roughskin newt (*Taricha granulosa*), the preoptic area and hypothalamus were found to contain estrogen receptors positive cells (Davis and Moore, 1996). Clearly, cyp19a1/aromatase and estrogen receptor expression studies revealed an overlapping distribution in specific brain areas, notably in the preoptic area and in the caudal hypothalamus. The apparent neuroanatomical distribution of aromatase and estrogen receptors also reveals an intriguing degree of consistency among vertebrates and suggests that estrogen signaling is involved in controlling similar brain functions and behaviors through evolution. It will be interesting to understand the roles of aromatase and estrogens in constitutive and regenerative neurogenesis in amphibians.

HYPOTHESIS AND PERSPECTIVES FOR ESTROGEN INVOLVEMENT IN NEUROGENESIS REGARDING NEW ADVANCES IN ZEBRAFISH AND NEW LITERATURE IN MAMMALS

In the past few years, constitutive and regenerative neurogenesis has been extensively studied in adult zebrafish, allowing the scientific community to reach a better understanding of neurogenic events. Here, we decided to highlight future directions that would be interesting to investigate for a better understanding of the roles of estrogens in neurogenic processes.

Recent data have shown that estrogens upregulate the cxcr4 receptor and its chemokine cxcl12 (Sdf1) in various cell-types in mammals (Boudot et al., 2011; Li et al., 2013). This chemokine/receptor couple is implicated in constitutive and regenerative neurogenesis, increasing neural stem cell survival (Zhu et al., 2012), regulating neuronal migration and differentiation (Cui et al., 2013; Yang et al., 2013), and acting on RGC morphology and cell fate in mammals (Mithal et al., 2013). Given these data and the recently described *cxcr4a/cxcl12b* expression in RGCs of adult zebrafish (Diotel et al., 2010b), it will be interesting to further investigate the roles of cxcr4a/cxcl12b in neurogenesis and its potential estrogenic regulation in zebrafish.

Recently, inflammation appeared as a positive regulator of neuronal regeneration in the zebrafish CNS. Indeed, acute inflammation initiates brain regeneration by increasing the neurogenic activity of RGCs (Kyritsis et al., 2012), notably as stab-wounded fish treated with the anti-inflammatory dexamethasone display a significant decrease in reactive neurogenesis (Kyritsis et al., 2012). However, the question of how inflammation enhances neurogenesis either by amplifying some signaling pathways or by initiating new and specific regenerative programs remains open. Given the complex pro- and anti-inflammatory effects of estrogens in mammals (Straub, 2007), notably during neuroinflammation and neurodegeneration (Spence and Voskuhl, 2012; Vegeto et al., 2001), it suggests that estrogens could participate in the regulation of inflammatory signals triggering injury-induced neurogenesis in zebrafish.

Another interesting direction to further explore is the link between notch and estrogen signaling. The notch pathway is well known for its involvement in the maintenance of neural progenitors and for regulating embryonic and adult neurogenesis (Imayoshi et al., 2010). In mice, recent evidence indicates that notch and estradiol signaling interact in the hippocampus, estradiol reducing notch signaling and finally favoring neuritogenesis (Arevalo et al., 2011; Wei et al., 2012). In zebrafish, notch signaling was recently described for its roles in constitutive and regenerative neurogenesis (Chapouton et al., 2010; Kishimoto et al., 2012; Rothenaigner et al., 2011). In the adult zebrafish, around 90% of proliferating RGCs express *notch1a*, *notch1b* and *notch3* while the other neural progenitors of the brain express these receptors at different levels and proportions (de Oliveira-Carlos et al., 2013). Taken together these data suggest that estradiol and notch signaling could interact for regulating neurogenic processes in fish.

Finally, there are accumulating data showing key roles for short (19–25 nucleotides) noncoding RNA called the microRNAs (miRNAs) in a wide range of physiological processes. miRNAs act by base-pairing with the 3′ untranslated region of target mRNAs, leading to translation blockade or transcript degradation. Recently, new studies documented their roles in neurogenesis in both mammals and zebrafish (Coolen et al., 2013; Ji et al., 2013; Lopez-Ramirez and Nicoli, 2013). Interestingly, some miRNAs appeared to be regulated by estrogens in mammals and zebrafish (Cohen and Smith, 2014; Klinge, 2009, 2012). Consequently, it would be interesting to study (1) how estrogen signaling could modulate miRNA expression and, reciprocally, (2) how miRNAs could regulate estrogen signaling in zebrafish during constitutive and reparative neurogenesis. However, miRNA targets and miRNA regulation are still poorly understood.

CONCLUSION

In vertebrates, there is increasing evidence showing that aromatase and estrogens are strongly implicated in constitutive and regenerative neurogenesis during development and adulthood. They notably act on neuroprotection, brain plasticity, spatial and working memory as well as behavior. For all these reasons, aromatase and estrogens are interesting candidates from a therapeutic point of view for fighting brain strokes and neurodegenerative diseases. However, we are still far from understanding their precise effects, given the different properties exhibited among estrogens, their pleiotropic effects, their complex signaling, and also their multifaceted combinatory effects with other steroids. A better understanding of these mechanisms is crucial for the development of new therapeutics, and also given the accumulating data showing how some endocrine disruptor chemicals could disturb estrogen signaling. In this chapter, we highlighted the power of zebrafish for studying the roles of estrogens and aromatase on adult neurogenesis and brain repair mechanisms. Altogether, these data suggest similarities but also major differences between fish and other vertebrates with respect to estrogen production and functions in adult and reparative neurogenesis (Figure 5.4). The rise of the zebrafish model will be reinforced in the next few years given its relevant and conserved physiological properties, and its utility for rapid drug screening, notably for assessing combined effects of estrogenic chemicals (Petersen et al., 2013), and the debate concerning the use of animals in laboratories.

References

Adolf, B., Chapouton, P., Lam, C.S., Topp, S., Tannhauser, B., Strahle, U., et al., 2006. Conserved and acquired features of adult neurogenesis in the zebrafish telencephalon. Dev. Biol. 295, 278–293.

Akatsuka, N., Komatsuzaki, E., Ishikawa, A., Suzuki, I., Yamane, N., Miyata, S., 2005. Expression of the gonadal p450 aromatase gene of *Xenopus* and characterization of the 5′-flanking region of the aromatase gene. J. Steroid Biochem. Mol. Biol. 96, 45–50.

Ampatzis, K., Makantasi, P., Dermon, C.R., 2012. Cell proliferation pattern in adult zebrafish forebrain is sexually dimorphic. Neuroscience. 226, 367–381.

Arevalo, M.A., Ruiz-Palmero, I., Scerbo, M.J., Acaz-Fonseca, E., Cambiasso, M.J., Garcia-Segura, L.M., 2011. Molecular mechanisms involved in the regulation of neuritogenesis by estradiol: recent advances. J. Steroid Biochem. Mol. Biol.

Azcoitia, I., Sierra, A., Veiga, S., Garcia-Segura, L.M., 2003. Aromatase expression by reactive astroglia is neuroprotective. Ann. N. Y. Acad. Sci. 1007, 298–305.

Balthazart, J., Ball, G.F., 1998. New insights into the regulation and function of brain estrogen synthase (aromatase). Trends Neurosci. 21, 243–249.

Barha, C.K., Galea, L.A., 2010. Influence of different estrogens on neuroplasticity and cognition in the hippocampus. Biochim. Biophys. Acta. 1800, 1056–1067.

Barha, C.K., Ishrat, T., Epp, J.R., Galea, L.A., Stein, D.G., 2011. Progesterone treatment normalizes the levels of cell proliferation and cell death in the dentate gyrus of the hippocampus after traumatic brain injury. Exp. Neurol. 231 (1), 72–81.

Baumgart, E.V., Barbosa, J.S., Bally-Cuif, L., Gotz, M., Ninkovic, J., 2012. Stab wound injury of the zebrafish telencephalon: a model for comparative analysis of reactive gliosis. Glia. 60, 343–357.

Becker, C.G., Becker, T., 2008. Adult zebrafish as a model for successful central nervous system regeneration. Restor. Neurol. Neurosci. 26, 71–80.

Bentivoglio, M., Mazzarello, P., 1999. The history of radial glia. Brain Res. Bull. 49, 305–315.

Boudot, A., Kerdivel, G., Habauzit, D., Eeckhoute, J., Le Dily, F., Flouriot, G., et al., 2011. Differential estrogen-regulation of CXCL12 chemokine receptors, CXCR4 and CXCR7, contributes to the growth effect of estrogens in breast cancer cells. PLoS One. 6, e20898.

Bowers, J.M., Waddell, J., McCarthy, M.M., 2010. A developmental sex difference in hippocampal neurogenesis is mediated by endogenous oestradiol. Biol. Sex Differ. 1, 8.

Brinton, R.D., 2009. Estrogen-induced plasticity from cells to circuits: predictions for cognitive function. Trends Pharmacol. Sci. 30, 212–222.

Brock, O., Keller, M., Veyrac, A., Douhard, Q., Bakker, J., 2010. Short term treatment with estradiol decreases the rate of newly generated cells in the subventricular zone and main olfactory bulb of adult female mice. Neuroscience. 166, 368–376.

Brunne, B., Zhao, S., Derouiche, A., Herz, J., May, P., Frotscher, M., et al., 2010. Origin, maturation, and astroglial transformation of secondary radial glial cells in the developing dentate gyrus. Glia. 58, 1553–1569.

Burrone, L., Santillo, A., Pinelli, C., Baccari, G.C., Di Fiore, M.M., 2012. Induced synthesis of P450 aromatase and 17beta-estradiol by D-aspartate in frog brain. J. Exp. Biol. 215, 3559–3565.

Callard, G., Schlinger, B., Pasmanik, M., 1990. Nonmammalian vertebrate models in studies of brain-steroid interactions. J. Exp. Zool. Suppl. 4, 6–16.

Callard, G.V., Petro, Z., Ryan, K.J., 1978. Phylogenetic distribution of aromatase and other androgen-converting enzymes in the central nervous system. Endocrinology. 103, 2283–2290.

Chang, X., Kobayashi, T., Senthilkumaran, B., Kobayashi-Kajura, H., Sudhakumari, C.C., Nagahama, Y., 2005. Two types of aromatase with different encoding genes, tissue distribution and developmental expression in Nile tilapia (*Oreochromis niloticus*). Gen. Comp. Endocrinol. 141, 101–115.

Chapouton, P., Jagasia, R., Bally-Cuif, L., 2007. Adult neurogenesis in non-mammalian vertebrates. Bioessays. 29, 745–757.

Chapouton, P., Skupien, P., Hesl, B., Coolen, M., Moore, J.C., Madelaine, R., et al., 2010. Notch activity levels control the balance between quiescence and recruitment of adult neural stem cells. J. Neurosci. 30, 7961–7974.

Chen, S.X., Bogerd, J., Garcia-Lopez, A., de Jonge, H., de Waal, P.P., Hong, W.S., et al., 2010. Molecular cloning and functional characterization of a zebrafish nuclear progesterone receptor. Biol. Reprod. 82, 171–181.

Cohen, A., Smith, Y., 2014. Estrogen regulation of microRNAs, target genes, and microRNA expression associated with vitellogenesis in the zebrafish. Zebrafish. 11, 462–478.

Cohen, R.E., Wade, J., 2011. Aromatase mRNA in the brain of adult green anole lizards: effects of sex and season. J. Neuroendocrinol. 23, 254–260.

Coolen, M., Katz, S., Bally-Cuif, L., 2013. miR-9: a versatile regulator of neurogenesis. Front. Cell. Neurosci. 7, 220.

Coumailleau, P., Kah, O., 2014. Cyp19a1 (aromatase) expression in the Xenopus brain at different developmental stages. J. Neuroendocrinol. 26, 226–236.

Cui, L., Qu, H., Xiao, T., Zhao, M., Jolkkonen, J., Zhao, C., 2013. Stromal cell-derived factor-1 and its receptor CXCR4 in adult neurogenesis after cerebral ischemia. Restor. Neurol. Neurosci. 31, 239–251.

D'Amico, L.A., Boujard, D., Coumailleau, P., 2011. Proliferation, migration and differentiation in juvenile and adult *Xenopus laevis* brains. Brain Res. 1405, 31–48.

D'Amico, L.A., Boujard, D., Coumailleau, P., 2013. The neurogenic factor NeuroD1 is expressed in post-mitotic cells during juvenile and adult *Xenopus* neurogenesis and not in progenitor or radial glial cells. PLoS One. 8, e66487.

Davis, G.A., Moore, F.L., 1996. Neuroanatomical distribution of androgen and estrogen receptor-immunoreactive cells in the brain of the male roughskin newt. J. Comp. Neurol. 372, 294–308.

de Oliveira-Carlos, V., Ganz, J., Hans, S., Kaslin, J., Brand, M., 2013. Notch receptor expression in neurogenic regions of the adult zebrafish brain. PLoS One. 8, e73384.

Diotel, N., Le Page, Y., Mouriec, K., Tong, S.K., Pellegrini, E., Vaillant, C., et al., 2010a. Aromatase in the brain of teleost fish: expression, regulation and putative functions. Front. Neuroendocrinol. 31, 172–192.

Diotel, N., Vaillant, C., Gueguen, M.M., Mironov, S., Anglade, I., Servili, A., et al., 2010b. Cxcr4 and Cxcl12 expression in radial glial cells of the brain of adult zebrafish. J. Comp. Neurol. 518, 4855–4876.

Diotel, N., Do Rego, J.L., Anglade, I., Vaillant, C., Pellegrini, E., Gueguen, M.M., et al., 2011a. Activity and expression of steroidogenic enzymes in the brain of adult zebrafish. Eur. J. Neurosci. 34, 45–56.

Diotel, N., Do Rego, J.L., Anglade, I., Vaillant, C., Pellegrini, E., Vaudry, H., et al., 2011b. The brain of teleost fish, a source, and a target of sexual steroids. Front. Neurosci. 5, 137.

Diotel, N., Servili, A., Gueguen, M.M., Mironov, S., Pellegrini, E., Vaillant, C., et al., 2011c. Nuclear progesterone receptors are up-regulated by estrogens in neurons and radial glial progenitors in the brain of zebrafish. PLoS One. 6, e28375.

Diotel, N., Vaillant, C., Gabbero, C., Mironov, S., Fostier, A., Gueguen, M.M., et al., 2013. Effects of estradiol in adult neurogenesis and brain repair in zebrafish. Horm. Behav. 63, 193–207.

Do Rego, J.L., Seong, J.Y., Burel, D., Leprince, J., Luu-The, V., Tsutsui, K., et al., 2009. Neurosteroid biosynthesis: enzymatic pathways and neuroendocrine regulation by neurotransmitters and neuropeptides. Front. Neuroendocrinol. 30, 259–301.

Ekström, P., Johnsson, C.M., Ohlin, L.M., 2001. Ventricular proliferation zones in the brain of an adult teleost fish and their relation to neuromeres and migration (secondary matrix) zones. J. Comp. Neurol. 436, 92–110.

Forlano, P.M., Deitcher, D.L., Myers, D.A., Bass, A.H., 2001. Anatomical distribution and cellular basis for high levels of aromatase activity in the brain of teleost fish: aromatase enzyme and mRNA expression identify glia as source. J. Neurosci. 21, 8943–8955.

Garcia-Segura, L.M., Wozniak, A., Azcoitia, I., Rodriguez, J.R., Hutchison, R.E., Hutchison, J.B., 1999. Aromatase expression by astrocytes after brain injury: implications for local estrogen formation in brain repair. Neuroscience. 89, 567–578.

Gelinas, D., Callard, G.V., 1997. Immunolocalization of aromatase- and androgen receptor-positive neurons in the goldfish brain. Gen. Comp. Endocrinol. 106, 155–168.

Gorelick, D.A., Watson, W., Halpern, M.E., 2008. Androgen receptor gene expression in the developing and adult zebrafish brain. Dev. Dyn. 237, 2987–2995.

Goto-Kazeto, R., Kight, K.E., Zohar, Y., Place, A.R., Trant, J.M., 2004. Localization and expression of aromatase mRNA in adult zebrafish. Gen. Comp. Endocrinol. 139, 72–84.

Götz, M., Huttner, W.B., 2005. The cell biology of neurogenesis. Nat. Rev. Mol. Cell. Biol. 6, 777–788.

Grandel, H., Brand, M., 2013. Comparative aspects of adult neural stem cell activity in vertebrates. Dev. Genes Evol. 223, 131–147.

Grandel, H., Kaslin, J., Ganz, J., Wenzel, I., Brand, M., 2006. Neural stem cells and neurogenesis in the adult zebrafish brain: origin, proliferation dynamics, migration and cell fate. Dev. Biol. 295, 263–277.

Greytak, S.R., Champlin, D., Callard, G.V., 2005. Isolation and characterization of two cytochrome P450 aromatase forms in killifish (*Fundulus heteroclitus*): differential expression in fish from polluted and unpolluted environments. Aquat. Toxicol. 71, 371–389.

Guiguen, Y., Fostier, A., Piferrer, F., Chang, C.F., 2010. Ovarian aromatase and estrogens: a pivotal role for gonadal sex differentiation and sex change in fish. Gen. Comp. Endocrinol. 165, 352–366.

Hanna, R.N., Daly, S.C., Pang, Y., Anglade, I., Kah, O., Thomas, P., et al., 2010. Characterization and expression of the nuclear progestin receptor in zebrafish gonads and brain. Biol. Reprod. 82, 112–122.

Hanna, R.N., Zhu, Y., 2009. Expression of membrane progestin receptors in zebrafish (*Danio rerio*) oocytes, testis and pituitary. Gen. Comp. Endocrinol. 161, 153–157.

Hao, R., Bondesson, M., Singh, A.V., Riu, A., McCollum, C.W., Knudsen, T.B., et al., 2013. Identification of estrogen target genes during zebrafish embryonic development through transcriptomic analysis. PLoS One. 8, e79020.

Imayoshi, I., Sakamoto, M., Yamaguchi, M., Mori, K., Kageyama, R., 2010. Essential roles of Notch signaling in maintenance of neural stem cells in developing and adult brains. J. Neurosci. 30, 3489–3498.

Iwabuchi, J., Arai, K., Miyata, S., 2008. Isolation of novel isoforms of estrogen receptor genes from *Xenopus* gonad and brain. Zoolog. Sci. 25, 1227–1233.

Jeng, S.R., Yueh, W.S., Pen, Y.T., Gueguen, M.M., Pasquier, J., Dufour, S., et al., 2012. Expression of aromatase in radial glial cells in the brain of the Japanese eel provides insight into the evolution of the cyp191a gene in Actinopterygians. PLoS One. 7, e44750.

Ji, F., Lv, X., Jiao, J., 2013. The role of microRNAs in neural stem cells and neurogenesis. J. Genet. Genomics. 40, 61–66.

Jorgensen, A., Andersen, O., Bjerregaard, P., Rasmussen, L.J., 2007. Identification and characterisation of an androgen receptor from zebrafish *Danio rerio*. Comp. Biochem. Physiol. C. Toxicol. Pharmacol. 146, 561–568.

Kishimoto, N., Shimizu, K., Sawamoto, K., 2012. Neuronal regeneration in a zebrafish model of adult brain injury. Dis. Model. Mech. 5, 200–209.

Kizil, C., Kaslin, J., Kroehne, V., Brand, M., 2012. Adult neurogenesis and brain regeneration in zebrafish. Dev. Neurobiol. 72, 429–461.

Klinge, C.M., 2009. Estrogen regulation of microRNA expression. Curr. Genomics. 10, 169–183.

Klinge, C.M., 2012. miRNAs and estrogen action. Trends Endocrinol. Metab. 23, 223–233.

Kroehne, V., Freudenreich, D., Hans, S., Kaslin, J., Brand, M., 2011. Regeneration of the adult zebrafish brain from neurogenic radial glia-type progenitors. Development. 138, 4831–4841.

Kyritsis, N., Kizil, C., Zocher, S., Kroehne, V., Kaslin, J., Freudenreich, D., et al., 2012. Acute inflammation initiates the regenerative response in the adult zebrafish brain. Science. 338, 1353–1356.

Le Page, Y., Diotel, N., Vaillant, C., Pellegrini, E., Anglade, I., Merot, Y., et al., 2010. Aromatase, brain sexualization and plasticity: the fish paradigm. Eur. J. Neurosci. 32, 2105–2115.

Li, H., Liu, J., Ye, X., Zhang, X., Wang, Z., Chen, A., et al., 2013. 17beta-Estradiol enhances the recruitment of bone marrow-derived endothelial progenitor cells into infarcted myocardium by inducing CXCR4 expression. Int. J. Cardiol. 162, 100–106.

Lindsey, B.W., Tropepe, V., 2006. A comparative framework for understanding the biological principles of adult neurogenesis. Prog. Neurobiol. 80, 281–307.

Liu, X., Bolteus, A.J., Balkin, D.M., Henschel, O., Bordey, A., 2006. GFAP-expressing cells in the postnatal subventricular zone display a unique glial phenotype intermediate between radial glia and astrocytes. Glia. 54, 394–410.

Liu, X., Zhu, P., Sham, K.W., Yuen, J.M., Xie, C., Zhang, Y., et al., 2009. Identification of a membrane estrogen receptor in zebrafish with homology to mammalian GPER and its high expression in early germ cells of the testis. Biol. Reprod. 80, 1253–1261.

Lopez-Ramirez, M.A., Nicoli, S., 2013. Role of miRNAs and epigenetics in neural stem cell fate determination. Epigenetics.9.

Martinez-Cerdeño, V., Noctor, S.C., Kriegstein, A.R., 2006. Estradiol stimulates progenitor cell division in the ventricular and subventricular zones of the embryonic neocortex. Eur. J. Neurosci. 24, 3475–3488.

März, M., Chapouton, P., Diotel, N., Vaillant, C., Hesl, B., Takamiya, M., et al., 2010. Heterogeneity in progenitor cell subtypes in the ventricular zone of the zebrafish adult telencephalon. Glia. 58, 870–888.

März, M., Schmidt, R., Rastegar, S., Strahle, U., 2011. Regenerative response following stab injury in the adult zebrafish telencephalon. Dev. Dyn. 240, 2221–2231.

Mensah-Nyagan, A.G., Do-Rego, J.L., Beaujean, D., Luu-The, V., Pelletier, G., Vaudry, H., 1999. Neurosteroids: expression of steroidogenic enzymes and regulation of steroid biosynthesis in the central nervous system. Pharmacol. Rev. 51, 63–81.

Menuet, A., Pellegrini, E., Anglade, I., Blaise, O., Laudet, V., Kah, O., et al., 2002. Molecular characterization of three estrogen receptor forms in zebrafish: binding characteristics, transactivation properties, and tissue distributions. Biol. Reprod. 66, 1881–1892.

Menuet, A., Anglade, I., Le Guevel, R., Pellegrini, E., Pakdel, F., Kah, O., 2003. Distribution of aromatase mRNA and protein in the brain and pituitary of female rainbow trout: comparison with estrogen receptor alpha. J. Comp. Neurol. 462, 180–193.

Menuet, A., Pellegrini, E., Brion, F., Gueguen, M.M., Anglade, I., Pakdel, F., et al., 2005. Expression and estrogen-dependent regulation of the zebrafish brain aromatase gene. J. Comp. Neurol. 485, 304–320.

Mithal, D.S., Ren, D., Miller, R.J., 2013. CXCR4 signaling regulates radial glial morphology and cell fate during embryonic spinal cord development. Glia. 61, 1288–1305.

Morrell, J.I., Kelley, D.B., Pfaff, D.W., 1975. Autoradiographic localization of hormone-concentrating cells in the brain of an amphibian, *Xenopus laevis*. II. Estradiol. J. Comp. Neurol. 164, 63–77.

Mouriec, K., Lareyre, J.J., Tong, S.K., Le Page, Y., Vaillant, C., Pellegrini, E., et al., 2009. Early regulation of brain aromatase (cyp19a1b) by estrogen receptors during zebrafish development. Dev. Dyn. 238 (10), 2641–2651.

Nagahama, Y., 2005. Molecular mechanisms of sex determination and gonadal sex differentiation in fish. Fish Physiol. Biochem. 31, 105–109.

Nagarajan, G., Tsai, Y.J., Chen, C.Y., Chang, C.F., 2011. Developmental expression of genes involved in neural estrogen biosynthesis and signaling in the brain of the orange-spotted grouper *Epinephelus coioides* during gonadal sex differentiation. J. Steroid Biochem. Mol. Biol. 127, 155–166.

Okubo, K., Takeuchi, A., Chaube, R., Paul-Prasanth, B., Kanda, S., Oka, Y., et al., 2011. Sex differences in aromatase gene expression in the medaka brain. J. Neuroendocrinol. 23, 412–423.

Onteniente, B., Kimura, H., Maeda, T., 1983. Comparative study of the glial fibrillary acidic protein in vertebrates by PAP immunohistochemistry. J. Comp. Neurol. 215, 427–436.

Ormerod, B.K., Lee, T.T., Galea, L.A., 2004. Estradiol enhances neurogenesis in the dentate gyri of adult male meadow voles by increasing the survival of young granule neurons. Neuroscience. 128, 645–654.

Pasmanik, M., Callard, G.V., 1985. Aromatase and 5 alpha-reductase in the teleost brain, spinal cord, and pituitary gland. Gen. Comp. Endocrinol. 60, 244–251.

Pellegrini, E., Menuet, A., Lethimonier, C., Adrio, F., Gueguen, M.M., Tascon, C., et al., 2005. Relationships between aromatase and estrogen receptors in the brain of teleost fish. Gen. Comp. Endocrinol. 142, 60–66.

Pellegrini, E., Mouriec, K., Anglade, I., Menuet, A., Le Page, Y., Gueguen, M.M., et al., 2007. Identification of aromatase-positive radial glial cells as progenitor cells in the ventricular layer of the forebrain in zebrafish. J. Comp. Neurol. 501, 150–167.

Pellegrini, E., Vaillant, C., Diotel, N., Benquet, P., Brion, F., Kah, O., 2013. Expression, regulation and potential functions of aromatase in radial glial cells of the fish brain. In: Balthazart, J., Ball, G.F. (Eds.), Brain Aromatase, Estrogens, and Behavior. Oxford, University Press, Oxford; New York, p. 2013.

Pérez, M.R., Pellegrini, E., Cano-Nicolau, J., Gueguen, M.M., Menouer-Le Guillou, D., Merot, Y., et al., 2013. Relationships between radial glial progenitors and 5-HT neurons in the paraventricular organ of adult zebrafish—potential effects of serotonin on adult neurogenesis. Eur. J. Neurosci. 38, 3292–3301.

Petersen, K., Fetter, E., Kah, O., Brion, F., Scholz, S., Tollefsen, K.E., 2013. Transgenic (cyp19a1b-GFP) zebrafish embryos as a tool for assessing combined effects of oestrogenic chemicals. Aquat. Toxicol. 138–139, 88–97.

Peterson, R.S., Saldanha, C.J., Schlinger, B.A., 2001. Rapid upregulation of aromatase mRNA and protein following neural injury in the zebra finch (*Taeniopygia guttata*). J. Neuroendocrinol. 13, 317–323.

Peterson, R.S., Lee, D.W., Fernando, G., Schlinger, B.A., 2004. Radial glia express aromatase in the injured zebra finch brain. J. Comp. Neurol. 475, 261–269.

Peterson, R.S., Yarram, L., Schlinger, B.A., Saldanha, C.J., 2005. Aromatase is pre-synaptic and sexually dimorphic in the adult zebra finch brain. Proc. Biol. Sci. 272, 2089–2096.

Rakic, P., 1978. Neuronal migration and contact guidance in the primate telencephalon. Postgrad. Med. J. 54 (Suppl. 1), 25–40.

Rakic, P., 1990. Principles of neural cell migration. Experientia. 46, 882–891.

Ravi, V., Venkatesh, B., 2008. Rapidly evolving fish genomes and teleost diversity. Curr. Opin. Genet. Dev. 18, 544–550.

Rothenaigner, I., Krecsmarik, M., Hayes, J.A., Bahn, B., Lepier, A., Fortin, G., et al., 2011. Clonal analysis by distinct viral vectors identifies bona fide neural stem cells in the adult zebrafish telencephalon and characterizes their division properties and fate. Development. 138 (8), 1459–1469.

Sakamoto, H., Ukena, K., Tsutsui, K., 2001. Activity and localization of 3beta-hydroxysteroid dehydrogenase/Delta5-Delta4-isomerase in the zebrafish central nervous system. J. Comp. Neurol. 439, 291–305.

Santoriello, C., Zon, L.I., 2012. Hooked! Modeling human disease in zebrafish. J. Clin. Invest. 122, 2337–2343.

Schmidt, R., Strahle, U., Scholpp, S., 2013. Neurogenesis in zebrafish—from embryo to adult. Neural Dev. 8, 3.

Shi, Y., Liu, X., Zhu, P., Li, J., Sham, K.W., Cheng, S.H., et al., 2013. G-protein-coupled estrogen receptor 1 is involved in brain development during zebrafish (Danio rerio) embryogenesis. Biochem. Biophys. Res. Commun. 435, 21–27.

Simmons, A.M., Horowitz, S.S., Brown, R.A., 2008. Cell proliferation in the forebrain and midbrain of the adult bullfrog, *Rana catesbeiana*. Brain Behav. Evol. 71, 41–53.

Spence, R.D., Voskuhl, R.R., 2012. Neuroprotective effects of estrogens and androgens in CNS inflammation and neurodegeneration. Front. Neuroendocrinol. 33, 105–115.

Straub, R.H., 2007. The complex role of estrogens in inflammation. Endocr. Rev. 28, 521–574.

Strobl-Mazzulla, P.H., Moncaut, N.P., Lopez, G.C., Miranda, L.A., Canario, A.V., Somoza, G.M., 2005. Brain aromatase from pejerrey fish (*Odontesthes bonariensis*): cDNA cloning, tissue expression, and immunohistochemical localization. Gen. Comp. Endocrinol. 143, 21–32.

Strobl-Mazzulla, P.H., Nunez, A., Pellegrini, E., Gueguen, M.M., Kah, O., Somoza, G.M., 2010. Progenitor radial cells and neurogenesis in pejerrey fish forebrain. Brain Behav. Evol. 76, 20–31.

Takase, M., Iguchi, T., 2007. Molecular cloning of two isoforms of *Xenopus* (Silurana) *tropicalis* estrogen receptor mRNA and their expression during development. Biochim. Biophys. Acta. 1769, 172–181.

Takeuchi, A., Okubo, K., 2013. Post-proliferative immature radial glial cells female-specifically express aromatase in the medaka optic tectum. PLoS One. 8, e73663.

Tanaka, E.M., Ferretti, P., 2009. Considering the evolution of regeneration in the central nervous system. Nat. Rev. Neurosci. 10, 713–723.

Tong, S.K., Mouriec, K., Kuo, M.W., Pellegrini, E., Gueguen, M.M., Brion, F., et al., 2009. A cyp19a1b-gfp (aromatase B) transgenic zebrafish line that expresses GFP in radial glial cells. Genesis. 47, 67–73.

Vaudry, H., Do Rego, J.L., Burel, D., Luu-The, V., Pelletier, G., Vaudry, D., et al., 2011. Neurosteroid biosynthesis in the brain of amphibians. Front. Endocrinol. (Lausanne). 2, 79.

Vegeto, E., Bonincontro, C., Pollio, G., Sala, A., Viappiani, S., Nardi, F., et al., 2001. Estrogen prevents the lipopolysaccharide-induced inflammatory response in microglia. J. Neurosci. 21, 1809–1818.

Veyrac, A., Bakker, J., 2011. Postnatal and adult exposure to estradiol differentially influences adult neurogenesis in the main and accessory olfactory bulb of female mice. FASEB J. 25, 1048–1057.

Wang, J.M., Liu, L., Brinton, R.D., 2008. Estradiol-17beta-induced human neural progenitor cell proliferation is mediated by an estrogen receptor beta-phosphorylated extracellularly regulated kinase pathway. Endocrinology. 149, 208–218.

Wang, L., Andersson, S., Warner, M., Gustafsson, J.A., 2003. Estrogen receptor (ER)beta knockout mice reveal a role for ERbeta in migration of cortical neurons in the developing brain. Proc. Natl. Acad. Sci. USA 100, 703–708.

Wei, Y., Zhang, Z., Liao, H., Wu, L., Wu, X., Zhou, D., et al., 2012. Nuclear estrogen receptor-mediated Notch signaling and GPR30-mediated PI3K/AKT signaling in the regulation of endometrial cancer cell proliferation. Oncol. Rep. 27, 504–510.

Yang, S., Edman, L.C., Sanchez-Alcaniz, J.A., Fritz, N., Bonilla, S., Hecht, J., et al., 2013. Cxcl12/Cxcr4 signaling controls the migration and process orientation of A9–A10 dopaminergic neurons. Development. 140, 4554–4564.

Zhang, Z., Yang, R., Zhou, R., Li, L., Sokabe, M., Chen, L., 2010. Progesterone promotes the survival of newborn neurons in the dentate gyrus of adult male mice. Hippocampus. 20, 402–412.

Zhu, B., Xu, D., Deng, X., Chen, Q., Huang, Y., Peng, H., et al., 2012. CXCL12 enhances human neural progenitor cell survival through a CXCR7- and CXCR4-mediated endocytotic signaling pathway. Stem Cells. 30, 2571–2583.

Zupanc, G.K., 2001. Adult neurogenesis and neuronal regeneration in the central nervous system of teleost fish. Brain Behav. Evol. 58, 250–275.

Zupanc, G.K., 2009. Towards brain repair: insights from teleost fish. Semin. Cell Dev. Biol. 20, 683–690.

Zupanc, G.K., Clint, S.C., 2003. Potential role of radial glia in adult neurogenesis of teleost fish. Glia. 43, 77–86.

Zupanc, G.K., Horschke, I., 1995. Proliferation zones in the brain of adult gymnotiform fish: a quantitative mapping study. J. Comp. Neurol. 353, 213–233.

CHAPTER

6

Neuroprotection by Exogenous Estrogenic Compounds Following Traumatic Brain Injury

Ana Belen Lopez-Rodriguez[2,3], *Marco Ávila-Rodriguez*[1], *Nelson E. Vega-Vela*[1], *Francisco Capani*[4,5,6], *Janneth Gonzalez*[1], *Luis Miguel García-Segura*[2], *and George E. Barreto*[1]

[1]Departamento de Nutrición y Bioquímica, Facultad de Ciencias, Pontificia Universidad Javeriana, Bogotá, Colombia [2]Instituto Cajal, CSIC, Madrid, Spain [3]Departamento de Fisiología Animal (II), Facultad de Biología, Universidad Complutense de Madrid, Spain [4]Laboratorio de Citoarquitectura e Injuria Neuronal, Instituto de Investigaciones Cardiológicas "Prof. Dr. Alberto C. Taquini" (ININCA), UBA-CONICET, Buenos Aires, Argentina [5]Departamento de Biología, Universidad Argentina John F Kennedy, Buenos Aires, Argentina [6]Facultad de Psicología, Universidad Católica Argentina, Buenos Aires, Argentina

INTRODUCTION

Traumatic brain injury (TBI) is the leading cause of death and disability in young and adult people and it is characterized by injuries that range from mild to severe (Krause and Richards, 2014; WHO, 2006). TBI occurs when the brain is subjected to forces that induce focal and diffuse damage such as cerebral edema, vascular and axonal injury and neuronal cell death (Burda and Sofroniew, 2014; Krause and Richards, 2014). Subsequently, secondary events triggered after this primary injury due to the first impact could lead to death or cause permanent disabilities (Hyder et al., 2007).

K.A. Duncan (Ed): *Estrogen Effects on Traumatic Brain Injury.*
DOI: http://dx.doi.org/10.1016/B978-0-12-801479-0.00006-1

Many different molecular pathways are activated after TBI and involve a spatial-temporal pattern that starts from the focus of the damage and spreads to the surrounding areas due to diverse mechanisms such as cell death, inflammation, cell survival and injury resolution (Burda and Sofroniew, 2014; Hyder et al., 2007). Several types of cells participate in the progression of TBI and each of them contributes differently in the degree of severity and the resolution of the damage, which also depends on the affected area (Burda and Sofroniew, 2014). Many authors propose that there are three critical steps after TBI: 1) Cell death and inflammation; 2) Tissue replacement; and 3) Tissue remodeling (Burda and Sofroniew, 2014; Kendall and Giuseppina, 2013; Roth et al., 2014). At each time point, specific and different dynamic events converge in the brain in response to TBI. Figure 6.1 shows astrocytic hyperreactivity in the border of a lesion in response to brain injury from a stab wound animal model.

Due to the alarming increase of TBI cases, several treatments have emerged to counteract the most critical processes after injury and diminish the subsequent damage that results from a complex event of cellular and molecular mechanisms (especially during the first week after injury) (Kendall and Giuseppina, 2013). For example, the first-choice therapies are directed to reduce clot formation, brain edema and inflammation and to increase energy sources in the surrounding and affected areas (Giulia, 2014). Follow-up therapies are administered according to the degree of severity after stabilization of the critical process (i.e., first week of the lesion) and are aimed to control inflammation, glial scar formation, hypertrophic reactive gliosis and synapse and circuit remodeling of the perilesion perimeter (Burda and Sofroniew, 2014).

In the last 30 years a great number of compounds and therapeutic alternatives have been tested, unfortunately without any substantial improvement (Onose et al., 2011; Giulia, 2014), probably due to the large

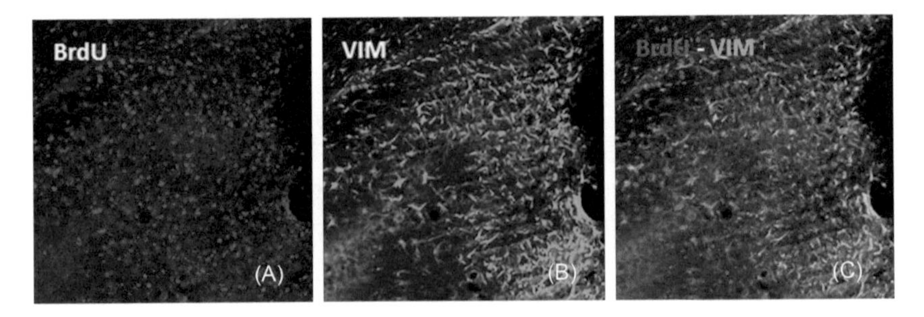

FIGURE 6.1 Panoramic view of the border of the lesion in CA1 showing representative examples of BrdU (A), vimentin (B) immunoreactivity in young rats after the administration of vehicle (corn oil). All figures are at the same magnification.

heterogeneity and variability of TBI lesions, which is a great disadvantage in finding effective therapies. However, neuroactive steroids present as good therapeutic candidates because of their pleiotropic neuroprotective effects related to their mechanism of action (Onose et al., 2011; Helen and Dietrich, 2001). Within neurosteroids, progesterone has been shown to improve the outcome in some clinical trials (Ajith et al., 1999; Courtney et al., 2006). Therefore, exogenous estrogenic compounds may be considered as a promising beneficial therapeutic approach (Arevalo et al., 2011). In this chapter, we will discuss and highlight several estrogenic compounds with neuroprotective potential for TBI. Within them, we will focus on selective estrogen receptor modulators (SERMs) and selective tissue estrogenic activity regulators (STEARs) that have been shown to exert neuroprotective properties in several *in vitro* and *in vivo* models (Kloosterboer, 2004; Lonard and Smith, 2002).

NEUROPROTECTIVE STEROIDS

In addition to the known effects of steroids on reproductive functions, a growing body of evidence has addressed a neuroprotective role of steroids in several types of injuries (Azcoitia et al., 2011; Barreto et al., 2007; Barreto et al., 2009; Barreto et al., 2014; Stein, 2008). In this context, both genomic and nongenomic mechanisms are involved in the protective effect of neurosteroids (Azcoitia et al., 2011). The genomic mechanisms involve the modulation of prosurvival genes (Strehlow et al., 2003), anti-inflammatory effects (Vegeto et al., 2008), anti-apoptotic functions and improved energy management (Garcia-Segura et al., 1998; Guo et al., 2012). These genomic mechanisms include the activation of Akt-PI3K and MAPK pathways that induce prosurvival mechanisms, upregulation of the anti-apoptotic molecule Bcl-2 and antioxidant enzymes SOD and GPx (Azcoitia et al., 2011; Simpkins and Dykens, 2008) and the downregulation of pro-inflammatory cytokines such as IL-1β, IL-6 and TNF-α (Rubio et al., 2011; Valles et al., 2010). The nongenomic effects of steroids are mainly associated with their antioxidant properties, especially because of the presence of an A-phenolic ring in their chemical structure (Aliev et al., 2008) that seems to exert its effects through an interaction with the estrogen-associated G-protein coupled receptor (GPR30) (Tang et al., 2014).

Previous evidence supports the beneficial effects of steroids following ischemia/reperfusion and traumatic brain injury in animal models (Liu et al., 2005; O'Connor et al., 2005) as well as steroid-demonstrated effectiveness in glucose deprivation and oxygen-glucose deprivation in *in vitro* models (Guo et al., 2012), suggesting these compounds as a promising therapy in central nervous system (CNS) injuries. Despite this evidence, the direct use of estrogens still represents a potential risk for human health as reported in some cases of hormonal replacement therapy in which women subjected to this treatment developed different types of cancer (uterine, ovarian or breast) and cerebral vascular diseases because of the genomic effects of estrogens (Colditz et al., 1995; Manolio et al., 1993). For this reason, it was imperative to develop compounds with similar protective action but without the side effects of estrogens, as we will explain in following subsections on SERMs and STEARs.

Estrogens, Progestagens and Androgens as Protective Compounds

Estrogen, progesterone and testosterone are reported as molecules involved in endocrine functions in many organs and tissues (Chedrese, 2009) that also exert protection after injury (Ajith et al., 1999; Azcoitia et al., 2011; Barreto et al., 2007). Following interaction with their specific receptor, each hormone triggers characteristic signaling pathways to stimulate differentiation, proliferation, growth and survival (Chedrese, 2009). The biosynthesis of estrogen, progesterone and testosterone takes place in endocrine organs, but it also occurs endogenously in brain tissue (Hojo et al., 2008). The estrogenic activity is mediated by the activation of alpha and beta estrogen receptors (ER) and G protein-coupled estrogen receptor (GPER) (Acaz-Fonseca et al., 2014). These receptors are activated by the endogenous ligands estrogen, estrone and estriol and also by other compounds such as alkylphenols, phthalates or polychlorinated biphenyls that present different binding affinities to these receptors (Pillon et al., 2005). The activation of ER—ERα and ERβ—implies their dimerization and nucleus translocation to regulate gene transcription by binding to the so-called estrogen response elements (ERE) (Cano et al., 2006). Within GPER receptors, GPR30 participates in rapid estrogen responses and it is located in the cell membrane (Cano et al., 2006). Although many of the activation pathways of estrogen are known, the complete molecular mechanisms of ERs are not completely elucidated yet, nor their possible location in other cells, raising new questions about their additional roles and functions (Otto et al., 2008; Yang et al., 2004).

Although progesterone (PROG) and estradiol (E2) present opposite endocrine functions, they share some prosurvival pathways and mechanisms (Singh and Su, 2013; Stein, 2008). Progesterone binds to its specific receptor (PR) regulating the transcription of target genes under the control of specific progesterone response elements (PRE) (Singh and Su, 2013). PROG also elicits fast responses through membrane PR (mPR) by mechanisms depending on protein kinase A (PKA) activation (Singh and Su, 2013) and their specific signaling pathways including MAPK, ERK1/2 and PI-3K/Akt. The activation of these signaling pathways may lead, among other effects, to the expression of prosurvival proteins such as Bcl-2 (Stein, 2008). Furthermore, PROG metabolites, such as allopregnanolone and 3α, 5α tetrahydroprogesterone, also mediate protective effects (Mahesh et al., 2006; Morali et al., 2011). It is known that allopregnanolone participates in mitochondrial functions by inhibiting ionic currents associated with the opening of the mitochondrial permeability transition pore (mtPTP) and so buffering the energy imbalance that occurs after any injury (Sayeed et al., 2009).

Similar to estrogens and progestagens, androgens also induce neuroprotection directly through the androgen receptor (AR) (Hammond et al., 2001). It is reported that testosterone and dihydrotestosterone reduce reactive gliosis by direct action through AR or by brain local conversion to estradiol (Barreto et al., 2007; Creta et al., 2010). After injury, reactive astrocytes express higher levels of aromatase enzyme, which converts testosterone to estradiol (Azcoitia et al., 2003), and this mechanism may partially explain the protective effects of testosterone on reactive microglia following penetrating brain injury (Barreto et al., 2007).

Dihydrotestosterone, a metabolic product of testosterone, is a potent agonist of AR and many of the biological effects of testosterone are mediated by this metabolite (Cheng et al., 2011). Barreto et al. (2007) propose that the early effects of dihydrotestosterone on reactive microglia are compatible with a differential role of testosterone metabolites in gliosis. It is suggested that androgens offer as much neuroprotection as 17β estradiol in a growth factor deprivation model that induces neuronal apoptosis (Hammond et al., 2001). Nonaromatizable molecules of androgens, such as mibolerone, also exert neuroprotection in a similar way to testosterone, suggesting that androgenic protective effects are not exclusively mediated by the aromatase conversion of testosterone to estradiol, but also because of their direct interaction with AR (Hammond et al., 2001), demonstrating that both genomic and nongenomic responses of ER, PR and AR elicit neuroprotective actions.

As previously mentioned, the direct use of estradiol, progesterone and testosterone as therapeutic compounds may have severe side effects, and due to their endocrine nature they also activate receptors in peripheral organs (Chedrese, 2009). The way to avoid these undesirable actions is using pharmacological compounds that take advantage of the estrogen, progesterone and testosterone neuroprotective properties without their detrimental effects (Arevalo et al., 2011). In this context, selective estrogen receptor modulators (SERMs) and selective tissue estrogenic activity regulators (STEARs) have been shown to be neuroprotective and to act as agonists or antagonists depending on the target organ (Arevalo et al., 2011; Kloosterboer, 2004). This tissue selective property of SERMs and STEARs is currently under investigation.

Selective Estrogen Receptor Modulators

The current knowledge about the structure and function of the ER has led to the concept that these receptors could trigger different pathways only by partial activation based on the idea that 1) ER have several molecular domains involved in their functions and 2) ER molecular domains interact within them to bind DNA elements, co-activators, co-repressors, chaperones and their respective ligands (Brzozowski et al., 1997; Cano et al., 2006). The establishment of this complex mechanism has been the result of years of evolution (Markov and Laudet, 2011) and to date it represents a key point in an attempt to completely elucidate the ER signaling mechanisms.

As other nuclear receptors, ER present a particular binding domain (LBD) to interact with their specific ligands (Cano et al., 2006). It is believed that the high or low affinity of the ligand with LBD plays a central role in the function of ER, because once the ligand binds LBD, conformational changes take place facilitating the interaction with co-activators, corepressors, DNA and other proteins (Cano et al., 2006). In this context, the conformational change is predetermined by the chemical nature of the ligand and its interaction with ER (McKenna et al., 1999). For example, the pharmacological development of compounds that emulate the binding activity of ligands such as estradiol, progesterone and testosterone could lead to different actions in ER, PR and AR in different tissues, depending on the cell context (Cano et al., 2006).

Tamoxifen is a clear example of a selective compound with estrogenic activity in liver, but anti-estrogenic activity in breast tissue (Shiau et al., 1998). These compounds are so-called selective estrogen receptor modulators (SERMs) and have been widely used in clinics for the treatment of breast cancer and as hormonal replacement therapy strategy (Vogelvang et al., 2004). SERMs are defined as compounds that are capable of binding ER and produce several responses, ranging from a pure estrogenic agonism to an anti-estrogen activity (Arevalo et al., 2011). Due to the diversity of cell context in complex organisms, the responses that SERMs may trigger are as diverse as their chemical nature. Previous works have shown that SERMs protect following spinal cord and traumatic brain injuries (Kokiko et al., 2006; Mosquera et al., 2014). Gonzales-Burgos et al. demonstrated that SERMs increase the number of dendritic spines of hippocampal neurons, suggesting a role in the processing of the information at a central level (González-Burgos et al., 2012). Additionally, raloxifene, a second-generation SERM, improved sensory motor and working memory deficits following TBI (Kokiko et al., 2006), suggesting that SERMs may act as potential therapeutic agents after CNS injury.

SERMs exert neuroprotection by activating nuclear factor-κB through PI3K-P38-ERK1/2 signaling (Arevalo et al., 2011), protein kinase C inhibitor, classical ER signaling (MAPK) (Arevalo et al., 2011), antioxidant enzyme expression such as manganese superoxide dismutase (MnSOD) (Wakade et al., 2008) or the endothelial nitric oxide synthase (eNOS) (Simoncini et al., 2002) and SERMs also induce upregulation of anti-apoptotic proteins such as Bcl-2 (Armagan et al., 2009). Altogether, the activation of these multifactorial protective signaling cascades can be critical for a proper therapeutic strategy in a high heterogeneous pathology, as is TBI.

Selective Tissue Estrogenic Activity Regulator

Women experience several hormonal changes during the menopause period that result in a variety of symptoms such as hot flashes, headache, insomnia, backache, mood/depression, dizziness and loss of libido. These downside effects are mostly related to low levels of estrogen and progesterone, which bring up these characteristic changes and challenges for women. Although the individual response to menopause varies considerably as a result of different factors such as genetic, cultural, lifestyle, socioeconomic, education and dietary factors (Sturdee, 2008), this decline deeply affects their quality of life. Hormonal replacement therapy (HRT) has been widely used for climacteric discomfort treatment, even though it has been reported that conventional HRT could have a relation to increases in cancer risk due to the use of estrogenic compounds together with progestin combinations (Kloosterboer, 2004). However, limited or nonconclusive evidence has been found to prove HRT benefits or harm in women's health due to a high variability in specific individual response (Barrett-Connor and Stuenkel, 2001). In the context of the risks and benefits associated with HRT, it has been established that hormone therapy with a combination of estrogen and progestagen increases the risk of different types of cancer, particularly breast and uterine cancers (Barrett-Connor and Stuenkel, 2001; Kloosterboer, 2004). This fact has encouraged many women to seek safer hormone therapy such as tibolone (Kloosterboer, 2004) or nonhormonal alternatives (e.g., botanical and dietary supplements) (Liu et al., 2001).

Tibolone has become a well-known treatment for climacteric symptoms (e.g., hot flashes, sweating and vaginal dryness) with fewer side effects than other HRT compounds, especially in women suffering low libido, persistent fatigue and blunted motivation (Genazzani et al., 2006; Gupta et al., 2013); in addition it has been used in the prevention of cardiovascular diseases, bone loss, osteoporosis in postmenopause

(Albertazzi et al., 1998; Campisi and Marengo, 2007). The clinical profile of this synthetic steroid exhibits advantages over estrogen plus progestagen combination treatment due to its unique characteristics that are mediated by the activation of the ER (Albertazzi et al., 1998; Reed and Kloosterboer, 2004). This steroid presents weak estrogenic, progestogenic and androgenic activities (Albertazzi et al., 1998; Escande et al., 2009; Kloosterboer, 2004) and a selective mode of action (Kloosterboer, 2001; Kloosterboer, 2004) that let us consider it as a new class of compound known as a selective tissue estrogenic activity regulator (STEAR).

STEARs are defined as compounds (e.g., synthetic or natural steroid) with estrogenic activity, tissue-selective mode of action and a particular metabolism that regulate ligand levels. According to Reed and Kloosterboer (2004), STEARs also show 1) estrogenic action on bone, vagina and brain without affecting breast or endometrium, 2) prereceptor modulation of ligand availability (e.g., regulated by enzymes and steroid metabolism) and 3) agonistic effects on estrogen-receptor alpha.

Escande et al. (2009) reported that tibolone acts as a pro-drug that has complex effects due to its particular mode of action on different steroid receptors. It is well reported that tibolone is metabolized in the body through two phase reactions to produce three different metabolites (Vos et al., 2002): two hydroxyl-metabolites (3-alpha-hydroxy- and 3-beta-hydroxy tibolone) as a result of 3-alpha and 3-beta-hydroxysteroid dehydrogenase activity, and one isomer (delta-4 tibolone) metabolized by 3-beta-hydroxysteroid dehydrogenase (Albertazzi et al., 1998; Escande et al., 2009; Kloosterboer, 2001). Each metabolite has different features. For example, tibolone and delta-4 tibolone are agonists for PR and AR (Escande et al., 2009), while 3-alpha and 3-beta-hydroxy metabolites are agonists for ER, but antagonists for PR and AR (Escande et al., 2009; Guzman et al., 2007). This tibolone-steroid receptor interaction and other regulatory mechanisms (e.g., local metabolism, steroid metabolizing enzymes, etc.) explain the tissue-selective effects of tibolone (Kloosterboer, 2004; Reed and Kloosterboer, 2004; Vos et al., 2002). Recently, neuroprotective action of tibolone has begun to be explored. Belenichev et al. (2012) used cortical neurons from neonatal rats to evaluate the neuroprotective activity of tibolone in a model of glutathione depletion that produces oxidative and nitrosative stresses and mitochondria dysfunction. These authors found that tibolone prevented mitochondria dysfunction and neuronal cell death. Similar neuroprotective results have been found in ovariectomized rats following cerebral ischemia injury (Wen-yan et al., 2009). Tibolone also presents anti-inflammatory effects that could help to prevent cardiovascular diseases in the long term (Campisi and Marengo, 2007; de Medeiros et al., 2012); however, more clinical studies and research are needed to determine their potential for medical applications.

TBI PROTECTION STRATEGIES

Strategies to protect and treat the brain after TBI are based on a complete understanding of the pathophysiological events and their consequences. Some of them stand out, such as cerebral edema, ischemia, hemorrhage, neuroinflammation, excitotoxicity or cell death. Furthermore, these physical events are accompanied by behavioral alterations and emotional symptoms such as impulsive behavior, apathy, lack of motivation, anxiety, depression, memory and sleeping problems or difficulties with language or making decisions (McAllister, 2008), which also represent a very disabling status for the patients and can persist well after the injury (Arciniegas et al., 2005).

The pathophysiology of TBI is still not well understood because it is a lesion that presents a highly variable, complex, and unpredictable prognosis. To understand these processes and their sequelae, many animal models have been developed. Roughly, they can be divided into penetrating (open models) or nonpenetrating injuries (closed models) (Figure 6.2). Open head injuries, considered as such when there is an opening of the dura, are quite uncommon, representing only 0.8–3% of patients (Masson et al., 2001; Myburgh et al., 2008; Wu et al., 2008). However, closed head models are closer to the injuries suffered in humans. Within the closed models, we can distinguish between controlled concussion, unconstrained impact acceleration and constrained impact acceleration models. Each of them produces a different kind of injuries that let us study specific parameters and TBI symptoms (Cernak, 2005). No single animal model reproduces completely the pathological changes after TBI.

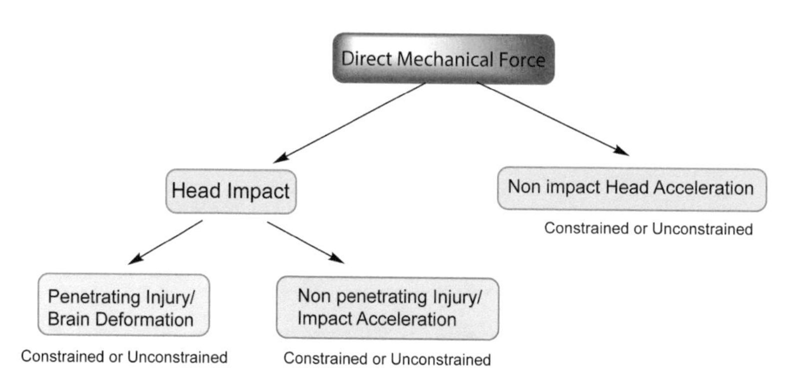

FIGURE 6.2 Schematic resume of *in vivo* experimental models of traumatic brain injury.

The vast majority of TBI preclinical studies of treatments are directed to reducing the first consequences of lesion: edema, neuroinflammation or cell death. For this aim, different drugs and therapies have been proposed. One possibility is trying to decrease neuroinflammation by inhibiting microglial activation, which is a key step in the development of secondary damage after trauma. For this purpose, many groups have used exogenous drugs such as minocycline, a derivative of the antibiotic tetracycline, which exerts anti-edematous, anti-inflammatory, anti-apoptotic and antioxidative effects up to three months after lesion (Siopi et al., 2011; Siopi et al., 2012), although its mechanism of action remains unclear (Plane et al., 2010).

Other approaches have taken advantage of the protective effect of the endocannabinoid system and used an endogenous cannabinoid as treatment after TBI in mice, demonstrating that the endocannabinoid 2-Arachidonoyl glycerol (2-AG) produces a reduction in brain edema, infarct volume and neuronal cell death as well as better recovery than control animals (Panikashvili et al., 2001). In addition, it has been shown that the use of antagonists of cannabinoid receptors worsens the symptoms of TBI and even blocks the effects of neuroprotective agents (Lopez-Rodriguez et al., 2013). Nonpharmacological studies tested the neuroprotective effect of hyperbaric oxygen (HBO) therapy showing a crucial role of the anti-inflammatory interleukine IL-10. HBO therapy reduces edema, lesion volume, inflammatory responses and improved motor and cognitive function as well as increased the expression of anti-apoptotic molecules like Bcl-2 (Chen et al., 2014).

Recent studies in animal models are starting to use stem cells in combination with growth factors, demonstrating an improvement of TBI outcome not only in histopathological features such as neurogenesis or cell loss but also in functional recovery such as motor function (Acosta et al., 2014; Torrente et al., 2013). Despite the different approximations, either using exogenous and endogenous compounds, nonpharmacological therapies or pioneer stem cell studies, there is no completely effective treatment for TBI yet.

ROLE OF NEUROACTIVE STEROIDS IN TBI

Due to the protective effects of neurosteroids in different type of injuries (see previous section titled "Neuroprotective Steroids"), they are promising candidates for TBI treatment, not only for the physiological impairments, but also to ameliorate some psychological alterations (Walf and Frye, 2006; Yang et al., 2014). Preclinical studies with gonadal hormones like testosterone, 17-beta estradiol, and progesterone, or with neurosteroids such as dehydroepiandrosterone, pregnenolone, and pregnenolone-sulfate, have demonstrated a downregulation of reactive glia after stab wound injury in rats (Barreto et al., 2007; Barreto et al., 2009; Barreto et al., 2014; Garcia-Estrada et al., 1999), demonstrating that their mechanism of action may be mediated by reactive astrocytes and reactive microglia. More recent studies suggest that some of these protective effects of 17-beta estradiol after TBI are mediated by the specific G protein-coupled estrogen receptor 1 (GPER), particularly those referred to neuronal survival, neuronal degeneration, and apoptotic cell death in the hippocampus (Day et al., 2013). This reveals a new potential therapeutic target in GPER, as well as the use of alternative activators or modulators of neurosteroid pathways. In a rough search on Google Scholar, we have estimated the number of cites that result for TBI + Estradiol (1.56%), TBI + Progesterone (1.72%), TBI + Tamoxifen (1.35%), TBI + Tibolone (0.079%) and TBI + Other hormones (6.03%). This gives us an idea of how recent the studies are on neuroprotection and the use of different hormones, SERMs and STEARs being considered for the treatment of TBI in different models (Figure 6.3).

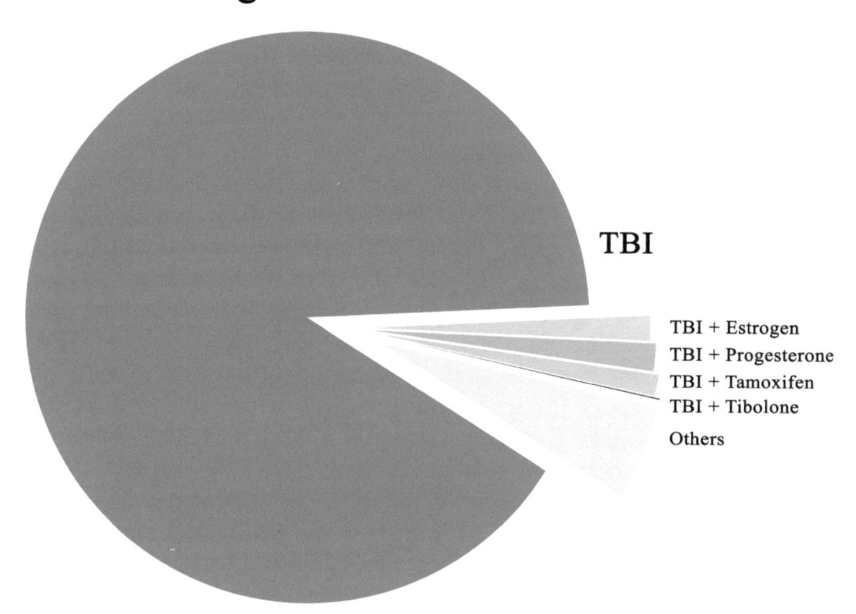

FIGURE 6.3 Pie chart representation of the number of citations in TBI studies.

In relation to new or alternative targets, SERMs and STEARs also appear as good candidates for neuroprotection following injury. There is currently scarce literature about the role of STEARs in neuroprotection. The most studied STEAR is tibolone, which possesses estrogenic, progestrogenic, and androgenic properties (Kloosterboer, 2004). The initial works with tibolone pertained to the prevention of bone loss and the treatment of osteoporosis in postmenopausal women, showing an advantage over estrogens and/or progestagen compounds in terms of having fewer side effects (Reed and Kloosterboer, 2004). In recent years, tibolone has been found to have neuroprotective traits, although it has not yet been used in TBI models. A recently published article demonstrated the neuroprotective effects of tibolone against oxidative stress in a rat model of ozone exposure. In this experiment, tibolone prevented lipid peroxidation, protein oxidation and neuronal death in hippocampus, as well as diminished the cognitive deficits in short- and long-term memory induced by ozone insult (Pinto-Almazan et al., 2014). Although much more evidence is needed in different animal models, tibolone appears as a possible neuroprotective treatment. SERMs have been deeply studied in a wide variety of injury models. For example, raloxifene has been tested in a rat TBI model showing improved functional recovery, facilitating the acquisition of working memory (Kokiko et al., 2006). Another SERM, tamoxifen, has been studied in a TBI model induced by fluid percussion in rats. This work revealed that tamoxifen decreased infarct size and motor impairments after TBI by a mechanism involving an increase in neuronal p-ERK1/2 expression, which would lead to higher levels of Bcl-2 and less neuronal apoptosis (Tsai et al., 2014).

Some of these compounds have been used in clinical trials, demonstrating a protective effect in patients with TBI. Currently, many clinical trials are recruiting participants to take part in studies with TBI and neurosteroids, especially those using progesterone. Shakeri et al. (2013) demonstrated that the use of progesterone in patients with diffuse axonal injury due to severe TBI improved the neurologic outcome up to 3 months after lesion, particularly those with $5 \leq$ Glasgow Coma Scale ≤ 8. Other clinical studies administered progesterone acutely and then once per 12 hours for 5 consecutive days to patients who arrived within 8 hours of injury with a Glasgow Coma Score ≤ 8. With this administration protocol, the results showed that TBI patients treated with progesterone hold improved neurologic outcomes for up to 6 months (Xiao et al., 2008). Although more preclinical and clinical studies are needed, it is clear that neurosteroids, selective estrogen modulators or activators are emerging as promising and effective neuroprotective drugs after TBI.

CONCLUSION AND PERSPECTIVES

Current evidence suggests that exogenous estrogenic compounds such as SERMs and STEARs function as neuroprotective agents for TBI because of their tissue-selective activity, multitarget protective properties and low cancer risk (Cano et al., 2006; Kloosterboer, 2004). In the past few decades, it has been demonstrated that the activation of estrogenic, progestogenic and androgenic receptors has beneficial effects. Therefore, those involved in the treatment of TBI should explore and investigate new pharmacological approaches to treat this pathological event, and to improve the critical recovery process up to several months or years after injury.

The pharmacological development of SERMs and STEARs opens a myriad of possibilities for the treatment of brain pathologies, due to their estrogenic protective advantages, focused on the nervous system without having systemic secondary effects (Arevalo et al., 2011; Barreto et al., 2009; Barreto et al., 2014; González-Burgos et al., 2012). It is imperative to refine and pursue more clinical evidence pertaining to SERMs and STEARs, as well as to develop new SERMs and STEARs that may improve the effects of estrogenic, progestogenic and androgenic neuroprotective activity.

Acknowledgements

This work was supported in part by grants PUJ IDs 4509, 5024 to GB and 5619 to JG.

References

Acaz-Fonseca, E., Sanchez-Gonzalez, R., Azcoitia, I., Arevalo, M.A., Garcia-Segura, L.M., 2014. Role of astrocytes in the neuroprotective actions of 17beta-estradiol and selective estrogen receptor modulators. Mol. Cell. Endocrinol. 389 (1–2), 48–57.

Acosta, S.A., Tajiri, N., Shinozuka, K., Ishikawa, H., Sanberg, P.R., Sanchez-Ramos, J., et al., 2014. Combination therapy of human umbilical cord blood cells and granulocyte colony stimulating factor reduces histopathological and motor impairments in an experimental model of chronic traumatic brain injury. PLoS One. 9 (3), e90953.

Ajith, J.T., Russ, P.N., Hiu, Q.P., Christopher, I.S., Michael, C., 1999. Progesterone is neuroprotective after acute experimental spinal cord trauma in rats. Spine. 24 (20), 2134.

Albertazzi, P., Di Micco, R., Zanardi, E., 1998. Tibolone: a review. Maturitas. 30 (3), 295–305.

Aliev, G., Obrenovich, M.E., Reddy, V.P., Shenk, J.C., Moreira, P.I., Nunomura, A., et al., 2008. Antioxidant therapy in Alzheimer's disease: theory and practice. Mini. Rev. Med. Chem. 8 (13), 1395–1406.

Arciniegas, D.B., Anderson, C.A., Topkoff, J., McAllister, T.W., 2005. Mild traumatic brain injury: a neuropsychiatric approach to diagnosis, evaluation, and treatment. Neuropsychiatr. Dis. Treat. 1 (4), 311–327.

Arevalo, M.A., Santos-Galindo, M., Lagunas, N., Azcoitia, I., Garcia-Segura, L.,M., 2011. Selective estrogen receptor modulators as brain therapeutic agents. J. Mol. Endocrinol. 46 (1), 9.

Armagan, G., Kanit, L., Terek, C.M., Sozmen, E.Y., Yalcin, A., 2009. The levels of glutathione and nitrite-nitrate and the expression of Bcl-2 mRNA in ovariectomized rats treated by raloxifene against kainic acid. Int. J. Neurosci. 119 (2), 227–239.

Azcoitia, I., Sierra, A., Veiga, S., Garcia-Segura, L.M., 2003. Aromatase expression by reactive astroglia is neuroprotective. Ann. N. Y. Acad. Sci. 1007, 298–305.

Azcoitia, I., Arevalo, M.A., De Nicola, A.F., Garcia-Segura, L.M., 2011. Neuroprotective actions of estradiol revisited. Trends. Endocrinol. Metab. 22 (12), 467–473.

Barreto, G., Veiga, S., Azcoitia, I., Garcia-Segura, L.M., Garcia-Ovejero, D., 2007. Testosterone decreases reactive astroglia and reactive microglia after brain injury in male rats: role of its metabolites, oestradiol and dihydrotestosterone. Eur. J. Neurosci. 25 (10), 3039–3046.

Barreto, G., Santos-Galindo, M., Diz-Chaves, Y., Pernia, O., Carrero, P., Azcoitia, I., et al., 2009. Selective estrogen receptor modulators decrease reactive astrogliosis in the injured brain: effects of aging and prolonged depletion of ovarian hormones. Endocrinology. 150 (11), 5010–5015.

Barreto, G.E., Santos-Galindo, M., Garcia-Segura, L.M., 2014. Selective Estrogen Receptor Modulators regulate reactive microglia after penetrating brain injury. Front. Aging Neurosci. 6.

Barrett-Connor, E., Stuenkel, C.A., 2001. Hormone replacement therapy (HRT)—risks and benefits. Int. J. Epidemiol. 30 (3), 423–426.

Belenichev, I.F., Odnokoz, O.V., Pavlov, S.V., Belenicheva, O.I., Polyakova, E.N., 2012. The neuroprotective activity of tamoxifen and tibolone during glutathione depletion in vitro. Neurochem. J. 6.

Brzozowski, A.M., Pike, A.C.W., Dauter, Z., Hubbard, R.E., Bonn, T., Engstrom, O., et al., 1997. Molecular basis of agonism and antagonism in the oestrogen receptor. Nature. 389 (6652), 753–758.

Burda, J.E., Sofroniew, M.V., 2014. Reactive gliosis and the multicellular response to CNS damage and disease. Neuron. 81 (2), 229–248.

Campisi, R., Marengo, F.D., 2007. Cardiovascular effects of tibolone: a selective tissue estrogenic activity regulator. Cardiovasc. Drug Rev. 25 (2), 132–145.

Cano, A., Alsina, J.C., Duenas-Diez, J.L., 2006. Selective Estrogen Receptor Modulators: A New Brand of Multitarget Drugs. Springer.

Cernak, I., 2005. Animal models of head trauma. NeuroRx. 2 (3), 410–422.

Chedrese, P.J., 2009. Reproductive Endocrinology: A Molecular Approach. Springer.

Chen, X., Duan, X.S., Xu, L.J., Zhao, J.J., She, Z.F., Chen, W.W., et al., 2014. Interleukin-10 mediates the neuroprotection of hyperbaric oxygen therapy against traumatic brain injury in mice. Neuroscience. 266, 235–243.

Cheng, J., Uchida, M., Zhang, W., Grafe, M.R., Herson, P.S., Hurn, P.D., 2011. Role of salt-induced kinase 1 in androgen neuroprotection against cerebral ischemia. J. Cereb. Blood Flow Metab. 31 (1), 339–350.

Colditz, G.A., Hankinson, S.E., Hunter, D.J., Willett, W.C., Manson, J.E., Stampfer, M.J., et al., 1995. The use of estrogens and progestins and the risk of breast cancer in postmenopausal women. N. Engl. J. Med. 332 (24), 1589–1593.

Courtney, L.R., April, P., Gloria, E.H., Anne, Z.M., Manda, S., Gary, F., 2006. Physiologic progesterone reduces mitochondrial dysfunction and hippocampal cell loss after traumatic brain injury in female rats. Exp. Neurol. 197 (1), 235243.

Creta, M., Riccio, R., Chiancone, F., Fusco, F., 2010. Androgens exert direct neuroprotective effects on the brain: a review of pre-clinical evidences. J. Androl. Sci. 17, 49–55.

Day, N.L., Floyd, C.L., D'Alessandro, T.L., Hubbard, W.J., Chaudry, I.H., 2013. 17beta-estradiol confers protection after traumatic brain injury in the rat and involves activation of G protein-coupled estrogen receptor 1. J. Neurotrauma. 30 (17), 1531–1541.

de Medeiros, A.R., Lamas, A.Z., Caliman, I.F., Dalpiaz, P.L., Firmes, L.B., de Abreu, G.R., et al., 2012. Tibolone has anti-inflammatory effects in estrogen-deficient female rats on the natriuretic peptide system and TNF-alpha. Regul. Pept. 179 (1–3), 55–60.

Escande, A., Servant, N., Rabenoelina, F., Auzou, G., Kloosterboer, H., Cavailles, V., et al., 2009. Regulation of activities of steroid hormone receptors by tibolone and its primary metabolites. J. Steroid Biochem. Mol. Biol. 116 (1–2), 8–14.

Garcia-Estrada, J., Luquin, S., Fernandez, A.M., Garcia-Segura, L.M., 1999. Dehydroepiandrosterone, pregnenolone and sex steroids down-regulate reactive astroglia in the male rat brain after a penetrating brain injury. Int. J. Dev. Neurosci. 17 (2), 145–151.

Garcia-Segura, L.M., Cardona-Gomez, P., Naftolin, F., Chowen, J.A., 1998. Estradiol upregulates Bcl-2 expression in adult brain neurons. Neuroreport. 9 (4), 593–597.

Genazzani, A.R., Pluchino, N., Bernardi, F., Centofanti, M., Luisi, M., 2006. Beneficial effect of tibolone on mood, cognition, well-being, and sexuality in menopausal women. Neuropsychiatr. Dis. Treat. 2 (3), 299–307.

Giulia, C., 2014. Therapeutic options to enhance coma arousal after traumatic brain injury: state of the art of current treatments to improve coma recovery. Br. J. Neurosurg. 28 (2), 187198.

González-Burgos, I., Rivera-Cervantes, M.C., Velázquez-Zamora, D.A., Feria-Velasco, A., Garcia-Segura, L.M., 2012. Selective estrogen receptor modulators regulate dendritic spine plasticity in the hippocampus of male rats. Neural. Plast. 2012, 309494.

Guo, J., Duckles, S.P., Weiss, J.H., Li, X., Krause, D.N., 2012. 17beta-Estradiol prevents cell death and mitochondrial dysfunction by an estrogen receptor-dependent mechanism in astrocytes after oxygen-glucose deprivation/reperfusion. Free Radic. Biol. Med. 52 (11–12), 2151–2160.

Gupta, B., Mittal, P., Khuteta, R., Bhargava, A., 2013. A Comparative Study of CEE, Tibolone, and DHEA as Hormone Replacement Therapy for Surgical Menopause. J. Obstet. Gynaecol. India. 63 (3), 194–198.

Guzman, C.B., Zhao, C., Deighton-Collins, S., Kleerekoper, M., Benjamins, J.A., Skafar, D. F., 2007. Agonist activity of the 3-hydroxy metabolites of tibolone through the oestrogen receptor in the mouse N20.1 oligodendrocyte cell line and normal human astrocytes. J. Neuroendocrinol. 19 (12), 958–965.

Hammond, J., Le, Q., Goodyer, C., Gelfand, M., Trifiro, M., LeBlanc, A., 2001. Testosterone-mediated neuroprotection through the androgen receptor in human primary neurons. J. Neurochem. 77 (5), 1319–1326.

Helen, M.B., Dietrich, W.D., 2001. Neuropathological protection after traumatic brain injury in intact female rats versus males or ovariectomized females. J. Neurotrauma. 18 (9), 891900.

Hojo, Y., Murakami, G., Mukai, H., Higo, S., Hatanaka, Y., Ogiue-Ikeda, M., et al., 2008. Estrogen synthesis in the brain—role in synaptic plasticity and memory. Mol. Cell. Endocrinol. 290 (1–2), 31–43.

Hyder, A.A., Wunderlich, C.A., Puvanachandra, P., Gururaj, G., Kobusingye, O.C., 2007. The impact of traumatic brain injuries: a global perspective. NeuroRehabilitation. 22 (5), 341–353.

Kendall, R.W., Giuseppina, T., 2013. Molecular mechanisms of cognitive dysfunction following traumatic brain injury. Front. Aging Neurosci. 5, 29.

Kloosterboer, H.J., 2001. Tibolone: a steroid with a tissue-specific mode of action. J. Steroid Biochem. Mol. Biol. 76 (1–5), 231–238.

Kloosterboer, H.J., 2004. Tissue-selectivity: the mechanism of action of tibolone. Maturitas. 48 (Suppl. 1), 40.

Kokiko, O.N., Murashov, A.K., Hoane, M.R., 2006. Administration of raloxifene reduces sensorimotor and working memory deficits following traumatic brain injury. Behav. Brain Res. 170 (2), 233–240.

Krause, M., Richards, S., 2014. Prevalence of traumatic brain injury and access to services in an undergraduate population: a pilot study. Brain Inj. 1–10.

Liu, J., Burdette, J.E., Xu, H., Gu, C., van Breemen, R.B., Bhat, K.P., et al., 2001. Evaluation of estrogenic activity of plant extracts for the potential treatment of menopausal symptoms. J. Agric. Food Chem. 49 (5), 2472–2479.

Liu, R., Wen, Y., Perez, E., Wang, X., Day, A.L., Simpkins, J.W., et al., 2005. 17beta-Estradiol attenuates blood–brain barrier disruption induced by cerebral ischemia-reperfusion injury in female rats. Brain Res. 1060 (1–2), 55–61.

Lonard, D.M., Smith, C.L., 2002. Molecular perspectives on selective estrogen receptor modulators (SERMs): progress in understanding their tissue-specific agonist and antagonist actions. Steroids. 67 (1), 15–24.

Lopez-Rodriguez, A.B., Siopi, E., Finn, D.P., Marchand-Leroux, C., Garcia-Segura, L.M., Jafarian-Tehrani, M., et al., 2013. CB1 and CB2 cannabinoid receptor antagonists prevent minocycline-induced neuroprotection following traumatic brain injury in mice. Cereb. Cortex. [epub before print].

Mahesh, V.B., Dhandapani, K.M., Brann, D.W., 2006. Role of astrocytes in reproduction and neuroprotection. Mol. Cell. Endocrinol. 246 (1–2), 1–9.

Manolio, T.A., Furberg, C.D., Shemanski, L., Psaty, B.M., O'Leary, D.H., Tracy, R.P., et al., 1993. Associations of postmenopausal estrogen use with cardiovascular disease and its risk factors in older women. The CHS Collaborative Research Group. Circulation. 88 (5 Pt 1), 2163–2171.

Markov, G.V., Laudet, V., 2011. Origin and evolution of the ligand-binding ability of nuclear receptors. Mol. Cell. Endocrinol. 334 (1–2), 21–30.

Masson, F., Thicoipe, M., Aye, P., Mokni, T., Senjean, P., Schmitt, V., et al., 2001. Epidemiology of severe brain injuries: a prospective population-based study. J. Trauma. 51 (3), 481–489.

McAllister, T.W., 2008. Neurobehavioral sequelae of traumatic brain injury: evaluation and management. World Psychiatry. 7 (1), 3–10.

McKenna, N.J., Lanz, R.B., O'Malley, B.W., 1999. Nuclear receptor coregulators: cellular and molecular biology. Endocr. Rev. 20 (3), 321–344.

Morali, G., Montes, P., Hernandez-Morales, L., Monfil, T., Espinosa-Garcia, C., Cervantes, M., 2011. Neuroprotective effects of progesterone and allopregnanolone on long-term cognitive outcome after global cerebral ischemia. Restor. Neurol. Neurosci. 29 (1), 1–15.

Mosquera, L., Colon, J.M., Santiago, J.M., Torrado, A.I., Melendez, M., Segarra, A.C., et al., 2014. Tamoxifen and estradiol improved locomotor function and increased spared tissue in rats after spinal cord injury: their antioxidant effect and role of estrogen receptor alpha. Brain Res. 1561, 11–22.

Myburgh, J.A., Cooper, D.J., Finfer, S.R., Venkatesh, B., Jones, D., et al., 2008. Epidemiology and 12-month outcomes from traumatic brain injury in Australia and New Zealand. J. Trauma. 64 (4), 854–862.

O'Connor, C.A., Cernak, I., Vink, R., 2005. Both estrogen and progesterone attenuate edema formation following diffuse traumatic brain injury in rats. Brain Res. 1062 (1–2), 171–174.

Onose, G., Daia-Chendreanu, C., Haras, M., Ciurea, A.V., Anghelescu, A., 2011. Traumatic brain injury: current endeavours and trends for neuroprotection and related recovery. Rom. Neurosurg. 15, 11–30.

Otto, C., Rohde-Schulz, B., Schwarz, G., Fuchs, I., Klewer, M., Brittain, D., et al., 2008. G protein-coupled receptor 30 localizes to the endoplasmic reticulum and is not activated by estradiol. Endocrinology. 149 (10), 4846–4856.

Panikashvili, D., Simeonidou, C., Ben-Shabat, S., Hanus, L., Breuer, A., Mechoulam, R., et al., 2001. An endogenous cannabinoid (2-AG) is neuroprotective after brain injury. Nature. 413 (6855), 527–531.

Pillon, A., Boussioux, A.M., Escande, A., Ait-Aissa, S., Gomez, E., Fenet, H., et al., 2005. Binding of estrogenic compounds to recombinant estrogen receptor-alpha: application to environmental analysis. Environ. Health Perspect. 113 (3), 278–284.

Pinto-Almazan, R., Rivas-Arancibia, S., Farfan-Garcia, E.D., Rodriguez-Martinez, E., Guerra-Araiza, C., 2014. Neuroprotective effects of tibolone against oxidative stress induced by ozone exposure. Rev. Neurol. 58 (10), 441–448.

Plane, J.M., Shen, Y., Pleasure, D.E., Deng, W., 2010. Prospects for minocycline neuroprotection. Arch. Neurol. 67 (12), 1442–1448.

Reed, M.J., Kloosterboer, H.J., 2004. Tibolone: a selective tissue estrogenic activity regulator (STEAR). Maturitas. 48 (Suppl. 1), S4–6.

Roth, T.L., Nayak, D., Atanasijevic, T., Koretsky, A.P., Latour, L.L., McGavern, D.B., 2014. Transcranial amelioration of inflammation and cell death after brain injury. Nature. 505 (7482), 223–228.

Rubio, N., Cerciat, M., Unkila, M., Garcia-Segura, L.M., Arevalo, M.A., 2011. An in vitro experimental model of neuroinflammation: the induction of interleukin-6 in murine astrocytes infected with Theiler's murine encephalomyelitis virus, and its inhibition by oestrogenic receptor modulators. Immunology. 133 (3), 360–369.

Sayeed, I., Parvez, S., Wali, B., Siemen, D., Stein, D.G., 2009. Direct inhibition of the mitochondrial permeability transition pore: a possible mechanism for better neuroprotective effects of allopregnanolone over progesterone. Brain Res. 1263, 165–173.

Shakeri, M., Boustani, M.R., Pak, N., Panahi, F., Salehpour, F., Lotfinia, I., et al., 2013. Effect of progesterone administration on prognosis of patients with diffuse axonal injury due to severe head trauma. Clin. Neurol. Neurosurg. 115 (10), 2019–2022.

Shiau, A.K., Barstad, D., Loria, P.M., Cheng, L., Kushner, P.J., Agard, D.A., et al., 1998. The structural basis of estrogen receptor/coactivator recognition and the antagonism of this interaction by tamoxifen. Cell. 95 (7), 927–937.

Simoncini, T., Genazzani, A.R., Liao, J.K., 2002. Nongenomic mechanisms of endothelial nitric oxide synthase activation by the selective estrogen receptor modulator raloxifene. Circulation. 105 (11), 1368–1373.

Simpkins, J., Dykens, J., 2008. Mitochondrial mechanisms of estrogen neuroprotection. Brain Res. Rev. 57 (2), 421–430.

Singh, M., Su, C., 2013. Progesterone and neuroprotection. Horm. Behav. 63 (2), 284–290.

Siopi, E., Cho, A.H., Homsi, S., Croci, N., Plotkine, M., Marchand-Leroux, C., et al., 2011. Minocycline restores sAPPalpha levels and reduces the late histopathological consequences of traumatic brain injury in mice. J. Neurotrauma. 28 (10), 2135–2143.

Siopi, E., Llufriu-Daben, G., Fanucchi, F., Plotkine, M., Marchand-Leroux, C., Jafarian-Tehrani, M., 2012. Evaluation of late cognitive impairment and anxiety states following traumatic brain injury in mice: the effect of minocycline. Neurosci. Lett. 511 (2), 110–115.

Stein, D.G., 2008. Progesterone exerts neuroprotective effects after brain injury. Brain Res. Rev. 57 (2), 386–397.

Strehlow, K., Rotter, S., Wassmann, S., Adam, O., Grohé, C., Laufs, K., et al., 2003. Modulation of antioxidant enzyme expression and function by estrogen. Circ. Res. 93 (2), 170–177.

Sturdee, D., 2008. Menopause. In: International Encyclopedia of Public Health International Encyclopedia of Public Health, pp. 335–343.

Tang, H., Zhang, Q., Yang, L., Dong, Y., Khan, M., Yang, F., et al., 2014. Reprint of "GPR30 mediates estrogen rapid signaling and neuroprotection". Mol. Cell. Endocrinol. 389 (1–2), 92–98.

Torrente, D., Avila, M., Cabezas, R., Morales, L., Gonzalez, J., Samudio, I., et al., 2013. Paracrine factors of human mesenchymal stem cells increase wound closure and reduce reactive oxygen species production in a traumatic brain injury in vitro model. Hum. Exp. Toxicol.

Tsai, Y.T., Wang, C.C., Leung, P.O., Lin, K.C., Chio, C.C., Hu, C.Y., et al., 2014. Extracellular signal-regulated kinase 1/2 is involved in a tamoxifen neuroprotective effect in a lateral fluid percussion injury rat model. J. Surg. Res. 189 (1), 106–116.

Valles, S.L., Dolz-Gaiton, P., Gambini, J., Borras, C., Lloret, A., Pallardo, F.V., et al., 2010. Estradiol or genistein prevent Alzheimer's disease-associated inflammation correlating with an increase PPAR gamma expression in cultured astrocytes. Brain Res. 1312, 138–144.

Vegeto, E., Benedusi, V., Maggi, A., 2008. Estrogen anti-inflammatory activity in brain: a therapeutic opportunity for menopause and neurodegenerative diseases. Front. Neuroendocrinol. 29 (4), 507–519.

Vogelvang, T.E., van der Mooren, M.J., Mijatovic, V., 2004. Hormone replacement therapy, selective estrogen receptor modulators, and tissue-specific compounds: cardiovascular effects and clinical implications. Treat. Endocrinol. 3 (2), 105–115.

Vos, R.M., Krebbers, S.F., Verhoeven, C.H., Delbressine, L.P., 2002. The in vivo human metabolism of tibolone. Drug Metab. Dispos. 30 (2), 106–112.

Wakade, C., Khan, M.M., De Sevilla, L.M., Zhang, Q.G., Mahesh, V.B., Brann, D.W., 2008. Tamoxifen neuroprotection in cerebral ischemia involves attenuation of kinase activation and superoxide production and potentiation of mitochondrial superoxide dismutase. Endocrinology. 149 (1), 367–379.

Walf, A.A., Frye, C.A., 2006. A review and update of mechanisms of estrogen in the hippocampus and amygdala for anxiety and depression behavior. Neuropsychopharmacology. 31 (6), 1097–1111.

Wen-yan, T., Hong-yi, Z., Li-kai, S., Wen-zeng, S., 2009. Effects of tibolone on apoptosis of neurons after cerebral ischemia-reperfusion injury in rats. Acad. J. Second Mil. Med. Univ. 30 (7), 790–792.

World Health Organization (WHO), 2006. The World Health Report 2006: Working Together for Health. WHO Press (WHO), Geneva.

Wu, X., Hu, J., Zhuo, L., Fu, C., Hui, G., Wang, Y., et al., 2008. Epidemiology of traumatic brain injury in eastern China, 2004: a prospective large case study. J. Trauma. 64 (5), 1313–1319.

Xiao, G., Wei, J., Yan, W., Wang, W., Lu, Z., 2008. Improved outcomes from the administration of progesterone for patients with acute severe traumatic brain injury: a randomized controlled trial. Crit. Care. 12 (2), R61.

Yang, F., Tao, J., Xu, L., Zhao, N., Chen, J., Chen, W., et al., 2014. Estradiol decreases rat depressive behavior by estrogen receptor beta but not alpha: no correlation with plasma corticosterone. Neuroreport. 25 (2), 100–104.

Yang, S.H., Liu, R., Perez, E.J., Wen, Y., Stevens Jr., S.M., Valencia, T., et al., 2004. Mitochondrial localization of estrogen receptor beta. Proc. Natl. Acad. Sci. U.S.A. 101 (12), 4130–4135.

CHAPTER

7

Neuroprotective Effects of Estrogen Following Neural Injury

Ashley B. Petrone[1,3,5], *Carolyn C. Rudy*[2], *Taura L. Barr*[3,5], *James W. Simpkins*[4,5], *and Miranda N. Reed*[1,2,5]

[1]Center for Neuroscience, West Virginia University, Morgantown, West Virginia, USA [2]Department of Psychology, West Virginia University, Morgantown, West Virginia, USA [3]School of Nursing, West Virginia University, Morgantown, West Virginia, USA [4]Department of Physiology and Pharmacology, School of Medicine, West Virginia University, Morgantown, West Virginia, USA [5]Center for Basic and Translational Stroke Research, West Virginia University Health Sciences Center, Morgantown, West Virginia, USA

INTRODUCTION

Approximately 50 million people suffer from neurodegenerative diseases annually in the United States, and nearly 25 percent are newly diagnosed. With almost 1 in 6 Americans suffering from a neurodegenerative disease, the annual financial cost to the US is several billion dollars in treatment and rehabilitation (Brown et al., 2005). The most common neurodegenerative disorders include Alzheimer's disease (AD), stroke, and traumatic brain injury (TBI). Despite the prevalence and enormous financial and physical burden of these conditions, treatment options are limited, and a large amount of research is focused on developing both preventative and acute treatments to limit neurodegeneration.

K.A. Duncan (Ed): Estrogen Effects on Traumatic Brain Injury.
DOI: http://dx.doi.org/10.1016/B978-0-12-801479-0.00007-3

While estrogens are most often associated with maintenance of female reproductive function, estrogens also elicit profound effects in the brain. Specifically, estrogens modulate a number of physiological processes that promote neuronal survival, as well as inhibit responses that contribute to cell loss. This chapter describes the common molecular mechanisms underlying neurodegeneration in AD, stroke, and TBI and provides evidence to support the use of estrogens in the treatment or prevention of these conditions.

MOLECULAR MECHANISM OF NEURODEGENERATION AND APOPTOSIS

Neurodegenerative diseases are marked by neuronal death that occurs either rapidly or gradually over time. While rapid neuronal cell loss may result in immediate symptoms and functional deficits, gradual cell loss accounts for a progressive worsening of the disease over time, as in AD. Regardless of the period of time over which neurodegeneration occurs, apoptosis is a common mechanism that mediates cell loss. Molecules that promote neuronal survival by increasing production of neuronal growth factors, reducing oxidative stress, and inhibiting apoptosis may have therapeutic potential that spans multiple neurodegenerative diseases, including TBI, AD, and stroke. While apoptosis is necessary in early stages of neuronal development, the death of mature neurons results in functional deficits in the damaged area. The symptoms of each disease may differ or overlap depending upon the specific populations of neurons affected. For example, a TBI damaging the motor cortex will likely result in motor impairment, whereas AD, which is characterized by loss of neurons in the hippocampus, is characterized by memory loss and cognitive decline, while motor function remains intact in the early stages of AD.

There are numerous signals that can trigger apoptosis, including oxygen deprivation, decreased metabolic rate, i.e., low adenosine triphosphate (ATP) production, oxidative stress, glutamate excitotoxicity, and loss of neurotrophic factors, such as brain-derived growth factor (BDNF) and nerve growth factor (NGF). Several of these signals increase calcium influx into neurons, and excessive intracellular calcium is a key mediator of apoptotic pathways in a cell, in many cases due to mitochondrial dysfunction (Figure 7.1). This calcium-induced ion imbalance activates the pro-apoptotic mitochondrial proteins, Bax and Bad, and while the exact downstream mechanism of Bax and Bad is unknown, Bax and Bad interact with other effector proteins in the cell to disrupt mitochondrial membrane potential. A disturbance in membrane potential triggers the opening of the mitochondrial permeability transition pore (MPTP) and release of cytochrome C into the cytoplasm, where cytochrome C activates pro-apoptotic caspase cascades (see Mattson, 2000 for review). Caspases

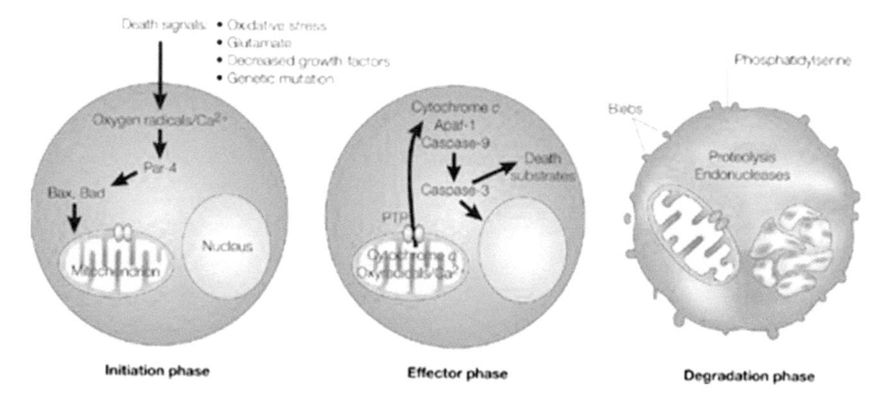

FIGURE 7.1 **Stages of apoptosis.** During the initiation phase of apoptosis, several signals can trigger an intracellular cascade of events that may involve increases in levels of free radicals and Ca^2 and translocation of pro-apoptotic Bcl-2 family members, Bax and Bad, to the mitochondrial membrane. Caspases can also act early in the cell death process before, or independently of, mitochondrial changes. The effector phase of apoptosis involves increased mitochondrial Ca^{2+}, the formation of permeability transition pores (PTP) in the mitochondrial membrane, and release of cytochrome *c* into the cytosol. Cytochrome *c* forms a complex with apoptotic protease-activating factor 1 (Apaf-1) and caspase-9. Activated caspase-9, in turn, activates caspase-3, which begins the degradation phase of apoptosis. *Reprinted with permission from Mattson, 2000.*

are universal apoptotic proteins present in virtually all cell types, including neurons and glial cells. One example of an apoptotic caspase cascade involves caspases 9 and 3. Cytochrome C binds caspase 9, with the aid of the protein protease-activating factor 1 (APAF1), the APAF1/caspase 9 complex activates the effector caspase, caspase 3, and caspase 3 stimulates DNA degradation and cell degeneration. Dysfunctional mitochondria also release free radicals into the cell that further increases oxidative stress and lipid peroxidation (see Mattson, 2000 for review).

In addition, neurons express several anti-apoptotic proteins, including the bcl-2 family members, bcl-2 and bcl-xL. Also, neurotrophic factors, such as BDNF and NGF, can prevent neuronal apoptosis by inhibiting pro-apoptotic pathways, activating anti-apoptotic signaling pathways, and stimulating production of molecules and proteins that promote cell survival. For example, BDNF and NGF increase the production of antioxidant molecules, as well as proteins that regulate calcium influx and homeostasis (Tamatani et al., 1998).

ESTROGEN RECEPTORS: TYPES AND LOCALIZATION IN THE BRAIN

Estrogens are lipophilic steroid hormones classically associated with regulating female reproductive function. Estrone (E1), estradiol (E2), and estriol (E3) are the three forms of estrogen synthesized from cholesterol by the ovaries, and to a lesser extent, the adrenal glands. 17-beta-estradiol (17β-E2) is the most potent estrogen. Throughout the course of a woman's reproductive years, 17β-E2 is the most prominent estrogen in circulation compared to E1 and E3; however, during menopause, ovaries cease to produce 17β-E2 from E1, and E1 becomes the principal form of circulating estrogen in postmenopausal women. Estrogens are distributed via the blood to a variety of tissues, including the cardiovascular, immune, and central nervous systems (Gustafson, 2003), and due to their lipophilic nature, estrogens can easily diffuse across cellular membranes, as well as the blood–brain barrier, to elicit their effects.

One way estrogens can elicit their actions is through interaction with estrogen receptors (ER). Currently, there are three known subtypes of ERs: ERα, ERβ, and G-protein coupled receptor 30 (GPR30/GPER). Elwood Jensen discovered ERα in 1958, and the first ERα knockout (KO) mouse was generated in 1993. Prior to 1993, ERα was widely believed to be the sole mediator of estrogenic action; however, this dogma was challenged following the discovery that female ERαKO mice had impaired reproductive function; however, all other vital physiological processes were unimpaired. This led to the hypothesis that more than one ER exists, and the discovery of ERβ in 1996 and GPR30 in 1997 provided rationale for the hypothesis of differential action of estrogens in different tissues. GPR30 is localized to the plasma membrane and was identified following evidence showing estrogenic actions in cells lacking ERα and ERβ (Filardo et al., 2000).

In addition to reproductive tissues, ERs are found in the central nervous system. ERα is widely distributed in the rodent and mouse forebrain (Shughrue and Merchenthaler, 2003). A similar distribution exists in the human brain, where ERα is distributed throughout the forebrain, hypothalamus, and hippocampus, but not in the cerebellum (Shughrue and Merchenthaler, 2003). In both rodents and humans, overall ERβ expression is much lower than ERα in the cerebral cortex; however, compared to humans, the ratio of ERβ/ERα is greater in the rodent midbrain and ERβ is expressed weakly in the cerebellum (Shughrue and Merchenthaler, 2003). GRP30 is expressed in the forebrain, hypothalamus, brainstem and hippocampus in both mice and rats (Brailoiu et al., 2007; Hazell et al., 2009). However, there are many factors that can alter ER expression and/or localization in the brain, including sex, E2 levels, and age. For example, ERα and ERβ expression declines in the hippocampus with age (Ishunina et al., 2007). This altered distribution of ERs in the brain may account, in part, for the age-associated decline in memory, as well as the differential action of estrogens across different species.

APOPTOTIC AND NEUROPROTECTIVE GENES AND PATHWAYS MODULATED BY ESTROGEN

There are two ligand-dependent ER signaling pathways: the genomic (classical) pathway or the nongenomic pathway (Hall et al., 2001; Heldring et al., 2007). In the genomic pathway, both ERα and ERβ serve as ligand-activated transcription factors. Once an ER binds estrogen, the active ER can form a homodimer (ERα/ERα or ERβ/ERβ) or heterodimer (ERα/ERβ) and translocate into the nucleus. The ligand-bound ER dimer can bind to estrogen response elements (ERE) in the promoter sequence of target genes, and once bound, the ligand-bound ER can recruit transcription factors or other coregulatory proteins to the promoter. The pool of coregulator proteins and transcription factors present in a cell will dictate the specific genomic action of the ligand-bound ER. Several genes involved in neuronal survival are located downstream of a promoter containing an ERE, and in turn, 17β-E2 promotes transcription of several genes involved in neuronal survival. Following neuronal insult, 17β-E2 administration increases expression of several proteins involved in cell survival, including phosphoinositide 3-kinase (PI3K) (Wang et al., 2006), Akt (Choi et al., 2004; Wang et al., 2006), cyclic-AMP response element binding protein (CREB) (Choi et al., 2004), Bcl-2 (Alkayed et al., 2001; Choi et al., 2004; Dubal et al., 1999; Singer et al., 1998), Bcl-x (Stoltzner et al., 2001), superoxide dismutase (SOD) (Rao et al., 2011), protein phosphatase 2A (PP2A) (Sung et al., 2010), c-fos (Rau et al., 2003), and c-jun (Rau et al., 2003). In addition, 17β-E2 inhibits expression of pro-apoptotic proteins, including Fas, FADD (Jia et al., 2009), and Bax (Choi et al., 2004), thereby inhibiting cytochrome c release (Choi et al., 2004).

In addition to the genomic pathway, estrogens signal through a nongenomic mechanism that occurs within seconds to minutes. A ligand-bound ER dimer can remain in the cytoplasm (ERα or ERβ) or at the plasma membrane (GPR30) and function as a signaling molecule through activation of protein kinases and phosphatases. 17β-E2 promotes neuronal survival through activation of cell survival proteins, such as the mitogen-activated protein kinases (MAPK) (Raval et al., 2009; Singer et al., 1999) Akt, CREB binding protein (Choi et al., 2004; Singer et al., 1999), and calcium-calmodulin-dependent protein kinase (CAMKII) (Raval et al., 2009). 17β-E2 can also inhibit apoptosis through several mechanisms, including increasing caspase-12 activation (Crosby et al., 2007) and inhibiting caspase-3 and caspase-8 activity (Jia et al., 2009).

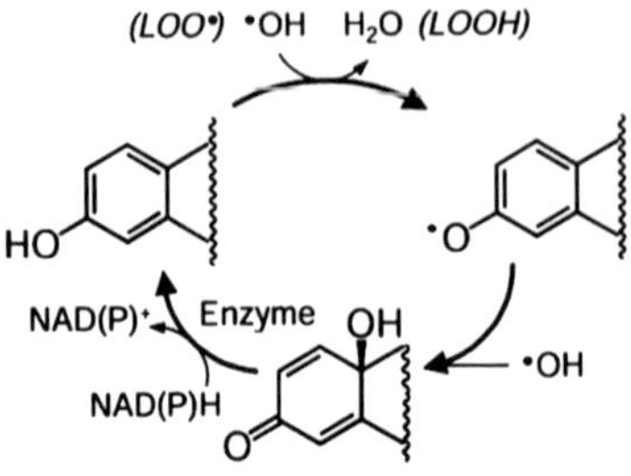

FIGURE 7.2 **Estrogen recycling.** Estrogens provide a chemical shield to neurons from free radical exposure. After the direct scavenging of radicals, the phenolic A-ring estrogen is reverted to an intermediate quinol molecule that is rapidly recycled to the parent estrogen through an enzyme-catalyzed reductive aromatization process. *Figure taken from Prokai et al., 2003. Reprinted with permission.*

ESTROGENS AS ANTIOXIDANTS

Oxidative stress and lipid peroxidation are common triggers of neurodegeneration in AD, stroke, and TBI. 17β-E2 is protective in several *in vitro* oxidative stress models (e.g., Barnham et al., 2004; Simpkins et al., 1997), including an oxidative stress model of Friedreich's ataxia (Richardson et al., 2011). Interestingly, the ataxia study is the first and only study to demonstrate that 17β-E2's protective action is not mediated by GPR30. The GPR30 agonist, G1, is unable to protect fibroblasts from oxidative stress, and the GPR30 antagonist, G15, does not antagonize the protective action of 17β-E2. These results indicate that the antioxidant activity of estrogen is mediated through ERα, ERβ, or ER-independent pathways.

Neuroprotective steroids must possess an intact phenolic A-ring (Behl et al., 1997; Green et al., 1997), and estrogens with more electron-donating substituent groups on the A ring are increasingly more powerful antioxidants. In accordance, any estrogen analogue that lacks a phenolic A-ring does not protect cells from oxidative stress or lipid peroxidation (Perez et al., 2005). Estrogens can donate an electron from their 3-OH group to neutralize extremely damaging hydroxyl and lipoxyl radicals (Prokai et al., 2003). The phenoxy radical produced is then rapidly reverted/recycled back into the original compound through enzymatic reduction (Figure 7.2). It is unknown how many redox cycles an estrogen can undergo, but this "chemical shield" likely underlies estrogen-mediated inhibition of oxidative stress and lipid peroxidation.

NEURAL INJURY AND THE IMMUNE RESPONSE

A balanced immune response is essential to limit tissue damage in neurodegenerative diseases. Damage-associated molecular patterns (DAMPs) released from dying neurons activate local microglia and promote recruitment of local and circulating leukocytes, and this milieu of immune cells secretes pro-inflammatory cytokines, such as IL-1β, IL-6, and TNFα (Kamel and Iadecola, 2012). Inflammation may also predispose individuals to neurodegenerative disease. For example, systemic inflammation is associated with an increased risk of ischemic stroke, with nearly 30 percent of strokes occurring in patients with a current or recent infection (Kamel and Iadecola, 2012). Immune cells and cytokines may be therapeutic targets in neurodegenerative diseases; however, further research is needed to elucidate what constitutes a "healthy" immune response versus a harmful immune response.

ESTROGEN AND THE IMMUNE RESPONSE

While a portion of estrogen's protective actions in brain injury are mediated by induction of neuroprotective pathways in the brain, estrogen is also a powerful immunomodulator. Because the immune response following injury dictates functional recovery and the extent of brain damage, estrogen may be dually protective by also mediating the immune response.

Innate Immunity

There is a rapid, local production of estrogen in the brain following brain injury, indicating that the hormone may be involved in an immediate physiological response to limit tissue damage (Garcia-Segura et al., 1999). This early production of estrogen occurs simultaneously with the innate arm of the immune response. Following injury, neutrophils rapidly migrate into the damaged brain region, and this process becomes more rapid and excessive as the blood–brain barrier becomes disrupted (Kamel and Iadecola, 2012). While adequate neutrophil infiltration is necessary to activate monocytes and macrophages to scavenge cellular debris from the site of injury, excessive neutrophil infiltration can exacerbate tissue damage. Estrogen inhibits the production of the neutrophil chemoattractants CXCL1, CXCL2 and CXCL3, thereby preventing an excessive neutrophil response (Nadkarni and McArthur, 2013). Aside from preventing an excessive neutrophil infiltration into the brain, estrogen also mediates the clearance of neutrophils from the brain after a balanced neutrophil response. Removal of apoptotic neutrophils by monocytes and macrophages is of critical importance for inflammatory resolution. Estrogen also induces an activated macrophage phenotype that secretes anti-inflammatory cytokines, such as IL-10 and TGFβ

(Iadecola and Anrather, 2011). In addition, estrogen prevents the production and secretion of pro-inflammatory cytokines, such as IL-1β, IL-6, and TNFα, through inhibition of NFkB signaling and transcription of pro-inflammatory genes (Nadkarni and McArthur, 2013).

Adaptive Immunity

Dendritic cells (DC) are often described as bridging the gap between the innate and adaptive immune system. Dendritic cells present antigens to T cells to stimulate the adaptive arm of the immune response, and estrogen promotes DC differentiation and MHC expression, thereby promoting a T cell response in the brain. Further, estrogen modulates the cytokine environment in the brain to control which subset of T cells will be recruited to the site of injury (Iadecola and Anrather, 2011). Cytotoxic T cell (CD8+) infiltration exacerbates tissue damage following brain injury, whereas T regulatory cells (Tregs) suppress pro-inflammatory responses and are often associated with decreased cell loss and better recovery (Nadkarni and McArthur, 2013). Estrogen inhibits production of pro-inflammatory, Th1 and Th17 cytokines, including IFNγ, TNFα and IL-17, as well as increases production of the anti-inflammatory cytokines, IL10 and TGFB, thereby generating a cytokine environment that favors a Treg response (Lakhan et al., 2009). Estrogen is also able to directly promote Treg proliferation through activation of the PI3K/AKT pathway (Iadecola and Anrather, 2011).

ESTROGEN AND TRAUMATIC BRAIN INJURY

TBI resulted in over two million visits to emergency departments in the United States in 2009 and is a significant public health concern at a cost of well over $60 billion each year (Coronado et al., 2012). Defined by the Centers for Disease Control and Prevention as blunt or penetrating trauma to the head associated with at least one of the following: alteration in consciousness, amnesia, neurologic or neuropsychiatric deviations, skull fracture(s), or intracranial lesion, TBI frequently occurs following transportation accidents, falls, and firearm-inflicted injuries (Coronado et al., 2011), but is of increasing concern as a common repercussion of warfare faced by military personnel (Huber et al., 2013). Because there are currently no available treatments that promote a functional recovery of motor function, memory, or cognitive ability, it is vital to investigate potential therapeutics that can alleviate the symptomology associated with brain trauma.

Estrogens, known to exhibit neuroprotective qualities, have garnered attention as possible targets for pharmacological intervention for TBI.

17β-E2 and estrone, which are antioxidant, anti-inflammatory, and anti-apoptotic agents (Gatson et al., 2012), show promising neuroprotective effects against the damage associated with traumatic brain injury (TBI). Despite the observation that the incidence of TBI is higher for men (651 per 100,000) than women (429 per 100,000), data on TBI outcome by sex is limited (Moore et al., 2010). However, evidence suggests that women exhibit better performance on post-TBI memory and executive functioning tests (Moore et al., 2010) and overall more favorable outcomes (Yeung et al., 2011) compared to men. Administration of estrogen preceding or immediately following a TBI in rats prevents apoptosis (Chen et al., 2009), and is thus being considered a viable treatment for patients with TBIs. The mechanism of estrogen's neuroprotection is not well characterized, but there is evidence for the role of estrogen in promoting neural regeneration following injury, including both TBI and cerebral ischemia (Chen et al., 2009).

Although *in vitro* models of TBI are less common than *in vivo* models, the few studies evaluating the effect of estrogen administration on *in vitro* models of TBI demonstrate the neuroprotective qualities of estrogen. Using a model of mechanical strain injury whereby the cell culture media is replaced with warm phosphate buffered saline and a 50-ms pressure pulse of compressed gas is introduced into the well, a rotational acceleration/deceleration brain injury can be emulated. With this model, the neuroprotective effects of estrogen were evidenced though decreased post-injury neuronal damage following the addition of 17β-E2 to neuronal-glial cultured cells (Lapanantasin et al., 2006). Protection from injury potentially results from the ability of cells to control intracellular calcium levels (Lapanantasin et al., 2006). Following a TBI, increased intracranial pressure compromises the blood–brain barrier, which then allows the flow of immune cells into the brain, resulting in oxidative injury and cell death (Nortje and Menon, 2004). It is possible that estrogens respond to these signals of cell stress to mediate damage. One proposed mechanism suggests that the secondary injury induced by increased intracranial pressure triggers survival mechanisms, including increased aromatase and estrogen levels (Gatson et al., 2011). When the precursor to estrone, androstenedione, is administered to primary cultured astrocytes before pressure-induced injury, there is a greater rate of conversion from androstenedione to estrone than in the noninjured group. When the cells were incubated with an aromatase inhibitor prior to the addition of androstenedione, estrone production was blocked, suggesting that estrogen levels are moderated by aromatase expression (Gatson et al., 2011). Results of *in vitro* studies demonstrating the neuroprotective value of estrogens have been corroborated by animal models of TBI.

Through a variety of *in vivo* models of TBI, estrogens exhibit neuroprotective qualities. Not only do low physiological levels of estrogen contribute to larger contusion sizes (Bramlett and Dietrich, 2001), but the administration of estrogens following a TBI can ameliorate negative outcomes. The effects of low estrogen levels can be studied through the use of males, or through the use of females that do not produce endogenous estrogens due to the removal of the female reproductive organs (i.e., ovariectomized). Following a TBI, males and ovariectomized females have greater levels of TUNEL positive cells and caspase-3, indicative of apoptosis and irreversible damage (Gatson et al., 2012; Li et al., 2007). When either estrone or Premarin®, an FDA-approved estrogen therapy, is administered prior to TBI, a significant reduction in two indicators of apoptosis, TUNEL positive cells and active caspase-3, in the cortex and hippocampus is observed (Chen et al., 2009; Soustiel et al., 2005). By decreasing apoptotic activity, there is a greater chance for a favorable outcome given that fewer cells have been irreparably damaged. Administration of Premarin also results in decreased blood glutamate concentrations in male rats that undergo a TBI induced by impact with a silicone-coated rod (Zlotnik et al., 2012). Abnormally high blood glutamate concentrations are associated with adverse outcomes for both TBI and stroke, and high blood glutamate levels are also correlated with poor neurological performance.

TBI results in a disruption of the blood–brain barrier that allows the infiltration of inflammatory agents into the brain. *In vivo* evidence corroborates the *in vitro* finding that estrogen may reduce damage incurred through immunological insult. Using a Marmarou TBI technique, a model of diffuse injury in which a weight is dropped onto the head, 17β-E2 administration attenuates the brain edema and blood–brain barrier disruption commonly observed following TBI (Asl et al., 2013). Further evidence for the role of estrogens in ameliorating inflammatory pathways comes from the ability of 17β-E2 to decrease neuroinflammation and apoptosis following a lateral fluid percussion model of TBI (Day et al., 2013) and burn-related injury. TBI and burn injury share a common mechanism in that both result in similar immediate brain changes, including the rapid elevation of pro-inflammatory cytokines, such as TNFα, IL-1β, and IL-6 (Gatson et al., 2009). Administration of 17β-E2 15 minutes after the burn injury significantly reduces levels of damaging cytokines and also blocks the activation of caspase-3, a key player in apoptosis. In addition, estradiol reduces reactive astroglia after brain injury, an effect mediated by the endocannabinoid system (López Rodríguez et al., 2011). In male rats that have undergone a TBI that emulates a stab wound to the brain, cannabinoid receptor antagonists administered just prior to injury blocked the protective ability of estradiol, indicating that cannabinoid receptors may contribute to the ability of estrogen to reduce resulting neuroinflammation (López Rodríguez et al., 2011).

ESTROGEN AND ALZHEIMER'S DISEASE

AD is the most common dementia, affecting well over 5 million individuals in the United States (Hebert et al., 2013) and is three times more prevalent in women than men (Alvarez-de-la-Rosa et al., 2005). The disease is characterized by three main hallmarks: insoluble plaques comprised of aggregated beta-amyloid protein, neurofibrillary tangles formed from hyperphosphorylated tau protein, and neuronal death. The disease progresses over the course of about 4 to 8 years during which time cognitive functioning declines steadily until death (Alzheimer's Association, 2012).

Pharmaceutical treatment of AD, which cannot halt or slow the progression of cognitive decline (Piau et al., 2010), is currently limited to two classes of drugs: noncompetitive N-methyl-D-aspartate (NMDA) receptor antagonists and reversible acetylcholinesterase inhibitors. As the American public ages, there is an urgent need to develop effective treatments that can address the enormous societal and economic burdens of this debilitating disease. Because women are significantly more likely to develop AD, it has been proposed that the postmenopausal decrease in estrogen contributes to molecular, cellular, and hormonal changes that induce pathological changes in the brain (Paganini-Hill and Henderson, 1994). The neuroprotective effect of estrogen has been the target of potential preventative therapies for AD (Simpkins et al., 2009). Both *in vitro* and *in vivo* evidence suggests that estrogen can prevent or reduce the pathological features of AD, including the formation of Aβ plaques and hyperphosphorylation of tau, as well as cognitive decline.

The use of estrogen *in vitro* has revealed neuroprotection against development of all three main pathological features of AD: Aβ plaques, neurofibrillary tangles, and neuronal death. *In vitro* application of estrogen also reduces Aβ-induced toxicity (Fitzpatrick et al., 2002; Marin et al., 2003), potentially by decreasing the production of Aβ or by increasing its degradation. In AD, Aβ plaques form when the amyloid precursor protein (APP) is cleaved into Aβ peptides by the proteolytic enzymes, α-, β-, and γ-secretase. Processing by the α-secretase pathway results in large N-terminal nonamyloidogenic soluble APP (sAPPα), whereas the β- and γ-secretase pathways lead to amyloidogenic Aβ that readily forms toxic aggregates. 17β-E2 decreases the formation of Aβ plaques by increasing α-secretase processing of APP into sAPPα, which does not readily form neurotoxic aggregates like those processed by either β- or γ-secretase (Manthey et al., 2001; Xu et al., 1998). In addition to preventing the formation of Aβ, 17β-E2 also encourages the degradation of Aβ peptides. By upregulating neprilysin (Liang et al., 2010) and insulin-degrading enzyme (Zhao et al., 2011), both enzymes that degrade Aβ, estrogen slows down the buildup of toxic Aβ fragments. Through slowing the production of Aβ and also increasing the breakdown, 17β-E2 attenuates Aβ-induced toxicity (Nilsen et al., 2006).

Use of estrogen can regulate the *in vitro* hyperphosphorylation of tau protein into neurofibrillary tangles; in human neuroblastoma SH-SY5Y cells treated with tangle-inducing okadaic acid, application of 17β-E2 prevents tau phosphorylation (Alvarez-de-la-Rosa et al., 2005; Zhang and Simpkins, 2010). By decreasing the activity of protein kinases responsible for phosphorylating pathological sites on the tau protein, 17β-E2 can also prevent tau phosphorylation in HEK293 cells expressing full-length tau protein. In particular, a decrease in tau phosphorylation can occur through a reduction in the overactivation of protein kinase A, which plays a crucial role in perpetuating tau pathology (Liu et al., 2008). Similarly, estrogen inactivates glycogen synthase kinase-3β (GSK-3β), another kinase known to phosphorylate tau at pathological sites, in an estrogen receptor-mediated manner (Goodenough et al., 2005). The ability of estrogen to prevent the induction of tau pathology suggests that estrogen may serve a neuroprotective role against the damage associated with neurofibrillary tangle formation.

Compelling evidence for the role of estrogen as a neuroprotective factor for AD has also been demonstrated in *in vivo* models. The depletion of estrogen results in increased beta-amyloid (Aβ) levels in guinea pigs (Carroll et al., 2007; Petanceska et al., 2000), and Aβ production and plaque formation in AD mouse models can be reversed with 17β-E2 administration (Zheng et al., 2002). 17β-E2 administered to ovariectomized mice that have AD-related mutations in Aβ (Levin-Allerhand et al., 2002), or both Aβ and tau mutations (Carroll et al., 2007), prevents the worsening of Aβ accumulation that accompanies the loss of estrogen, as well as the associated memory deficits (Carroll et al., 2007). Similarly, estrogen therapy significantly reduces the associated increases in Aβ levels (Levin-Allerhand et al., 2002; Xu et al., 2006).

With respect to tau pathology, the sudden loss of estrogen resulting from an ovariectomy is associated with tau accumulation and hyperphosphorylation. 17β-E2 administration reduces the abnormal tau hyperphosphorylation associated with two risk factors of AD: transient cerebral ischemia (Wen et al., 2004; Zhang et al., 2008) and Down syndrome (Hunter et al., 2004). One possible mechanism for the neuroprotective role of estrogens against tau pathology includes the prevention of signaling pathways that promote neurodegeneration. For instance, 17β-E2 prevents the induction of Dkk1, a neurodegenerative factor that serves as an antagonist for the Wnt/β-catenin signaling pathway (Zhang et al., 2008). In particular, the Wnt/β-catenin signaling pathway protects the CA1 region of the hippocampus from damage. Because the Wnt/β-catenin signaling pathway becomes activated from lack of Dkk1 antagonism, there is subsequent protection from ischemia-induced neuronal death in the hippocampus, a brain region particularly important for memory. Given that estrogens have been effective at alleviating multiple pathological hallmarks of AD, additional research is warranted to fully characterize the neuroprotective potential of estrogens for therapeutic intervention.

ESTROGEN AND ISCHEMIC STROKE

Stroke is a leading cause of long-term disability and the fourth leading cause of death in the United States following heart disease, cancer, and chronic lower respiratory disease. Approximately 795,000 strokes are reported each year in the US, 87% of which are ischemic strokes (Go et al., 2013). The direct medical cost associated with stroke in 2009 was approximately $22.8 billion, with an additional $13.8 billion in indirect costs associated with lost productivity, unemployment, rehabilitation, and follow-up care (Go et al., 2013).

Despite the significant physical and economic burdens of stroke, treatment options are limited. Tissue plasminogen activator (tPA) is a thrombolytic agent that breaks down blood clots to restore blood flow to the ischemic region of the brain and is the only FDA-approved drug available to treat ischemic stroke. Unfortunately, tPA is only effective if administered within 4½ hours of stroke onset. Many stroke patients do not reach a treatment facility within this time window, and thus are ineligible to receive tPA (Go et al., 2013). As of 2012, 114 drugs have entered clinical trials to evaluate their efficacy in ischemic stroke treatment; however, all have since failed to show efficacy in reducing ischemic damage (Lakhan et al., 2009). Given the incidence of ischemic stroke in the US and the limited therapeutic window of tPA, it is critical to develop a more effective treatment for ischemic stroke.

Women tend to have more severe strokes, more stroke deaths, and increased poststroke functional deficits than men (Appelros et al., 2009). Although lifetime risk of stroke is higher for men than for women, from ages 19–30, and again from ages 45–54, women have an increased stroke risk compared to men. One plausible explanation for the increased risk of stroke during these time periods is alterations in estrogen status. The first stroke surge is likely due to a high rate of childbirth over this age range. During pregnancy, maternal estrogen levels rise due to an increased estrogen production by the placenta. Following childbirth, estrogen levels decrease rapidly, while still remaining elevated compared to prepregnancy levels. These rapid decreases and changes in estrogen levels result in many health issues, including an increased risk of ischemic stroke (Koellhoffer and McCullough, 2013). The second stroke surge is likely due to the menopausal transition.

Typically occurring near age 50, menopause is the cessation of reproductive fertility in women due to a decreased production of circulating sex hormones, such as estrogen and progesterone (Ritzel et al., 2013). This transition into reproductive senescence and long-term estrogen deprivation is accompanied by symptoms that diminish a woman's quality of life, including hot flashes, night sweats, insomnia, weight gain, and osteoporosis. Women who undergo menopause before age 42 have a doubled lifetime risk of stroke compared to women undergoing menopause after 51 years of age (Go et al., 2013). Evidence also suggests that a decreased length of time between menarche and menopause, resulting in a decrease in overall estradiol exposure, increases the risk of ischemic stroke (De Leciñana et al., 2007).

Ischemic stroke is characterized by an abrupt deprivation of blood flow, oxygen, and nutrients to the brain that quickly leads to cell death in the core of the ischemic brain region. Initiation of the ischemic cascade begins within minutes of ischemia onset and involves several events, such as increased oxidative stress and mitochondrial dysfunction, resulting in apoptosis and a progressive loss of cells in the penumbra that may continue hours to days after stroke. Estrogen has been shown to mediate many events of the ischemic cascade, thereby providing a probable link to sex differences in stroke incidence and progression.

17β-E2 is neuroprotective in *in vivo* models of focal cerebral ischemia. Neuroprotection has been demonstrated in ovariectomized female mice, rodents, and gerbils (Dubal et al., 1998; Rusa et al., 1999; Shughrue and Merchenthaler, 2003; Simpkins et al., 1997; Yang et al., 2000; Zhang et al., 1998), in both young and middle-aged animals (Dubal and Wise, 2001). Gibson et al. (2006) compiled an extensive review of all experimental studies to date that have examined the effects of estrogen treatment on cerebral ischemia. Overall, estrogens reduce lesion volume in a dose-dependent manner when administered up to a week before or up to four hours after transient or permanent cerebral ischemia (Gibson et al., 2006). While most experimental studies demonstrate that both pre- and posttreatment with 17β-E2 reduces lesion size and infarct volume in models of focal cerebral ischemia, there are a limited number of studies that suggest no effect or a more negative outcome as a result of 17β-E2 treatment (Harukuni et al., 2001; Leon et al., 2012).

LIMITATIONS/FUTURE DIRECTIONS

Despite the prevailing amount of experimental evidence supporting the neuroprotective role of estrogens, the Women's Health Initiative (WHI) study was ended early because of findings indicating increased risks of cardiovascular disease, stroke, blood clots, breast cancer, and dementia for women on estrogen therapy. The WHI began in 1993 and consisted of two double-blind, placebo-controlled clinical trials to determine whether estrogen alone or estrogen given with progestin would reduce the number of cardiovascular events in postmenopausal women. The estrogen-alone trial enrolled 10,739 postmenopausal women between 50 and 79 years of age with prior hysterectomy. The WHI estrogen-alone trial was scheduled to end between October 2004 and March 2005; however, the study was terminated in February 2004 when conjugated equine estrogens did not affect the risk of cardiovascular disease in participants. Aside from failing to reduce the incidence of heart disease, the incidence of stroke was increased by 39 percent in the conjugated equine estrogens group when compared to placebo (Anderson et al., 2004).

Re-evaluation of the surprising findings of the WHI has generated several hypotheses to account for the increased incidence of stroke among women receiving estrogen therapy (Lobo, 2013). First, the increased risk of stroke reported by the WHI is of borderline significance. Eliminating data from subjects with comorbidities, such as obesity and hypertension, may render the increased risk insignificant. Second, the age of the subjects can confound the results of the WHI. While the overall risk of stroke significantly increased, the increased risk of stroke among younger women (<50 years of age) was negligible (Dubey et al., 2005; Grodstein et al., 2008). The average age of WHI participants was 63 years of age, approximately 13 years after the average onset of menopause (Anderson et al., 2004). Overall, 83 percent of WHI subjects were at least 5 years into menopause. The critical window hypothesis suggests that initiation of estrogen therapy near the onset of menopause in younger women has a higher benefit/risk ratio, whereas ET in older women who have been in menopause for an extended period of time may be ineffective or harmful (Coker et al., 2009; Grodstein et al., 2008). Additionally, the route of estrogen administration may impact the results of the WHI. In the WHI, estrogen was given orally, and data suggests that transdermal administration may be safer and more effective than oral administration by avoiding first-pass metabolism and bioactivation (Grodstein et al., 2008). It is also important to consider that in experimental studies, estrogens are generally administered via intravenous or subcutaneous injection, and this difference may shed some light on the discrepancy between experimental stroke models and clinical trials.

CONCLUSIONS

Minimizing the off-target effects associated with the deleterious outcomes of estrogen therapy, while identifying and targeting the sites responsible for estrogen's neuroprotective effects, remains a top priority for the field. Some of the deleterious effects resulting from estrogen therapy may be due to activation of ERs in peripheral tissue (e.g., breast and uterus). Use of compounds, such as nonfeminizing estrogens, which exhibit protective actions independent of activation of the known ERs, ERα, ERβ, and GPR30, may allow for an ET strategy with minimal side effects by avoiding activation of ERs in peripheral tissue. Similarly, identifying the parameters, such as duration, route of administration, and cyclic vs. tonic administration, by which ET confers protection is essential to making ET a viable option for neural injury.

Acknowledgements

The research described herein was supported in part by the following grants: P01 AG022550 (JWS), P01 AG027956 (JWS), a grant from the Robert Wood Johnson Nurse Faculty Scholars Program (TLB) and National Institute of General Medical Sciences, U54GM104942 (MNR), and the Alzheimer's Association, NIRG-12-242187 (MNR).

References

Alkayed, N.J., Goto, S., Sugo, N., Joh, H.D., Klaus, J., Crain, B.J., et al., 2001. Estrogen and Bcl-2: gene induction and effect of transgene in experimental stroke. J. Neurosci. 21 (19), 7543–7550.

Alvarez-de-la-Rosa, M., Silva, I., Nilsen, J., Pérez, M.M., García-Segura, L.M., Avila, J., et al., 2005. Estradiol prevents neural tau hyperphosphorylation characteristic of Alzheimer's disease. Ann. N. Y. Acad. Sci. 224, 210–224.

Alzheimer's Association, 2012. Alzheimer's disease facts and figures. Alzheimers Dement. 8 (2), 131–168.

Anderson, G.L., Limacher, M., Assaf, A.R., Bassford, T., Beresford, S.A.A., Black, H., et al., 2004. Effects of conjugated equine estrogen in postmenopausal women with hysterectomy: the Women's Health Initiative randomized controlled trial. JAMA. 291 (14), 1701–1712.

Appelros, P., Stegmayr, B., Terént, A., 2009. Sex differences in stroke epidemiology: a systematic review. Stroke. 40 (4), 1082–1090.

Asl, S.Z., Khaksari, M., Khachki, A.S., Shahrokhi, N., Nourizade, S., 2013. Contribution of estrogen receptors alpha and beta in the brain response to traumatic brain injury. J. Neurosurg. 119 (2), 353–361.

Barnham, K.J., Masters, C.L., Bush, A.I., 2004. Neurodegenerative diseases and oxidative stress. Nat. Rev. Drug Discov. 3 (3), 205–214.

Behl, C., Skutella, T., Lezoualc'h, F., Post, A., Widmann, M., Newton, C.J., et al., 1997. Neuroprotection against oxidative stress by estrogens: structure-activity relationship. Mol. Pharmacol. 51 (4), 535–541.

Brailoiu, E., Dun, S.L., Brailoiu, G.C., Mizuo, K., Sklar, L.A., Oprea, T.I., et al., 2007. Distribution and characterization of estrogen receptor G protein-coupled receptor 30 in the rat central nervous system. J. Endocrinol. 193 (2), 311–321.

Bramlett, H.M., Dietrich, W.D., 2001. Neuropathological protection after traumatic brain injury in intact female rats versus males or ovariectomized females. J. Neurotrauma. 18 (9), 891–900.

Brown, R.C., Lockwood, A.H., Sonawane, B.R., 2005. Neurodegenerative diseases: an overview of environmental risk factors. Environ. Health Perspect. 113 (9), 1250–1256.

Carroll, J.C., Rosario, E.R., Chang, L., Stanczyk, F.Z., Oddo, S., LaFerla, F.M., et al., 2007. Progesterone and estrogen regulate Alzheimer-like neuropathology in female 3xTg-AD mice. J. Neurosci. 27 (48), 13357–13365.

Chen, S.-H., Chang, C.-Y., Chang, H.-K., Chen, W.-C., Lin, M.-T., Wang, J.-J., et al., 2009. Premarin stimulates estrogen receptor-alpha to protect against traumatic brain injury in male rats. Crit. Care Med. 37 (12), 3097–3106.

Choi, Y.C., Lee, J.H., Hong, K.W., Lee, K.S., 2004. 17 Beta-estradiol prevents focal cerebral ischemic damages via activation of Akt and CREB in association with reduced PTEN phosphorylation in rats. Fundam. Clin. Pharmacol. 18 (5), 547–557.

Coker, L.H., Hogan, P.E., Bryan, N.R., Kuller, L.H., Margolis, K.L., Bettermann, K., et al., 2009. Postmenopausal hormone therapy and subclinical cerebrovascular disease: the WHIMS-MRI Study. Neurology. 72 (2), 125–134.

Coronado, V.G., Xu, L., Basavaraju, S.V., McGuire, L.C., Wald, M.M., Faul, M.D., et al., 2011. Surveillance for traumatic brain injury—related deaths in the United States. Surveill. Summ. 60 (SS05), 1–32.

Coronado, V.G., McGuire, L.C., Sarmiento, K., Bell, J., Lionbarger, M.R., Jones, C.D., et al., 2012. Trends in traumatic brain injury in the U.S. and the public health response: 1995–2009. J. Safety Res. 43 (4), 299–307.

Crosby, K.M., Connell, B.J., Saleh, T.M., 2007. Estrogen limits ischemic cell death by modulating caspase-12-mediated apoptotic pathways following middle cerebral artery occlusion. Neuroscience. 146 (4), 1524–1535.

Day, N.L., Floyd, C.L., D'Alessandro, T.L., Hubbard, W.J., Chaudry, I.H., 2013. 17β-estradiol confers protection after traumatic brain injury in the rat and involves activation of G protein-coupled estrogen receptor 1. J. Neurotrauma. 30 (17), 1531–1541.

De Leciñana, M.A., Egido, J.A., Fernández, C., Martínez-Vila, E., Santos, S., Morales, A., et al., 2007. PIVE Study Investigators of the Stroke Project of the Spanish Cerebrovascular Diseases Study Group. Risk of ischemic stroke and lifetime estrogen exposure. Neurology. 68 (1), 33–38.

Dubal, D.B., Wise, P.M., 2001. Neuroprotective effects of estradiol in middle-aged female rats. Endocrinology. 142 (1), 43–48.

Dubal, D.B., Kashon, M.L., Pettigrew, L.C., Ren, J.M., Finklestein, S.P., Rau, S.W., et al., 1998. Estradiol protects against ischemic injury. J. Cereb. Blood Flow Metab. 18 (11), 1253–1258.

Dubal, D.B., Shughrue, P.J., Wilson, M.E., Merchenthaler, I., Wise, P.M., 1999. Estradiol modulates bcl-2 in cerebral ischemia: a potential role for estrogen receptors. J. Neurosci. 19 (15), 6385–6393.

Dubey, R.K., Imthurn, B., Barton, M., Jackson, E.K., 2005. Vascular consequences of menopause and hormone therapy: importance of timing of treatment and type of estrogen. Cardiovasc. Res. 66 (2), 295–306.

Filardo, E.J., Quinn, J.A., Bland, K.I., Frackelton, A.R., 2000. Estrogen-induced activation of Erk-1 and Erk-2 requires the G protein-coupled receptor homolog, GPR30, and occurs via trans-activation of the epidermal growth factor receptor through release of HB-EGF. Mol. Endocrinol. 14 (10), 1649–1660.

Fitzpatrick, J.L., Mize, A.L., Wade, C.B., Harris, J.A., Shapiro, R.A., Dorsa, D.M., 2002. Estrogen-mediated neuroprotection against beta-amyloid toxicity requires expression of estrogen receptor alpha or beta and activation of the MAPK pathway. J. Neurochem. 82 (3), 674–682.

Garcia-Segura, L.M., Wozniak, A., Azcoitia, I., Rodriguez, J.R., Hutchison, R.E., Hutchison, J.B., 1999. Aromatase expression by astrocytes after brain injury: implications for local estrogen formation in brain repair. Neuroscience. 89 (2), 567–578.

Gatson, J.W., Maass, D.L., Simpkins, J.W., Idris, A.H., Minei, J.P., Wigginton, J.G., 2009. Estrogen treatment following severe burn injury reduces brain inflammation and apoptotic signaling. J. Neuroinflammation. 6, 30.

Gatson, J.W., Simpkins, J.W., Yi, K.D., Idris, A.H., Minei, J.P., Wigginton, J.G., 2011. Aromatase is increased in astrocytes in the presence of elevated pressure. Endocrinology. 152 (1), 207–213.
Gatson, J.W., Liu, M.-M., Abdelfattah, K., Wigginton, J.G., Smith, S., Wolf, S., et al., 2012. Estrone is neuroprotective in rats after traumatic brain injury. J. Neurotrauma. 29 (12), 2209–2219.
Gibson, C.L., Gray, L.J., Murphy, S.P., Bath, P.M.W., 2006. Estrogens and experimental ischemic stroke: a systematic review. J. Cereb. Blood Flow Metab. 26 (9), 1103–1113.
Go, A.S., Mozaffarian, D., Roger, V.L., Benjamin, E.J., Berry, J.D., Borden, W.B., et al., 2013. Heart disease and stroke statistics—2013 update: a report from the American Heart Association. Circulation. 127 (1), e6–e245.
Goodenough, S., Schleusner, D., Pietrzik, C., Skutella, T., Behl, C., 2005. Glycogen synthase kinase 3β links neuroprotection by 17β-estradiol to key Alzheimer processes. Neuroscience. 132 (3), 581–589.
Green, P.S., Gordon, K., Simpkins, J.W., 1997. Phenolic A ring requirement for the neuroprotective effects of steroids. J. Steroid Biochem. Mol. Biol. 63 (4–6), 229–235.
Grodstein, F., Manson, J.E., Stampfer, M.J., Rexrode, K., 2008. Postmenopausal hormone therapy and stroke: role of time since menopause and age at initiation of hormone therapy. Arch. Intern. Med. 168 (8), 861–866.
Gustafsson, J.-Å., 2003. What pharmacologists can learn from recent advances in estrogen signalling. Trends Pharmacol. Sci. 24 (9), 479–485.
Hall, J.M., Couse, J.F., Korach, K.S., 2001. The multifaceted mechanisms of estradiol and estrogen receptor signaling. J. Biol. Chem. 276 (40), 36869–36872.
Harukuni, I., Hurn, P.D., Crain, B.J., 2001. Deleterious effect of beta-estradiol in a rat model of transient forebrain ischemia. Brain Res. 900 (1), 137–142.
Hazell, G.G.J., Yao, S.T., Roper, J.A., Prossnitz, E.R., O'Carroll, A.-M., Lolait, S.J., 2009. Localisation of GPR30, a novel G protein-coupled oestrogen receptor, suggests multiple functions in rodent brain and peripheral tissues. J. Endocrinol. 202 (2), 223–236.
Hebert, L.E., Weuve, J., Scherr, P.A., Evans, D.A., 2013. Alzheimer disease in the United States (2010–2050) estimated using the 2010 census. Neurology. 80 (19), 1778–1783.
Heldring, N., Pike, A., Andersson, S., Matthews, J., Cheng, G., Hartman, J., et al., 2007. Estrogen receptors: how do they signal and what are their targets. Physiol. Rev. 87 (3), 905–931.
Huber, B.R., Meabon, J.S., Martin, T.J., Mourad, P.D., Bennett, R., Kraemer, B.C., et al., 2013. Blast exposure causes early and persistent aberrant phospho- and cleaved-tau expression in a murine model of mild blast-induced traumatic brain injury. J. Alzheimers Dis. 37, 309–323.
Hunter, C.L., Bimonte-Nelson, H.A., Nelson, M., Eckman, C.B., Granholm, A.-C., 2004. Behavioral and neurobiological markers of Alzheimer's disease in Ts65Dn mice: effects of estrogen. Neurobiol. Aging. 25 (7), 873–884.
Iadecola, C., Anrather, J., 2011. The immunology of stroke: from mechanisms to translation. Nat. Med. 17 (7), 796–808.
Ishunina, T.A., Fischer, D.F., Swaab, D.F., 2007. Estrogen receptor alpha and its splice variants in the hippocampus in aging and Alzheimer's disease. Neurobiol. Aging. 28 (11), 1670–1681.
Jia, J., Guan, D., Zhu, W., Alkayed, N.J., Wang, M.M., Hua, Z., et al., 2009. Estrogen inhibits Fas-mediated apoptosis in experimental stroke. Exp. Neurol. 215 (1), 48–52.
Kamel, H., Iadecola, C., 2012. Brain-immune interactions and ischemic stroke: clinical implications. Arch. Neurol. 69 (5), 576–581.
Koellhoffer, E.C., McCullough, L.D., 2013. The effects of estrogen in ischemic stroke. Transl. Stroke Res. 4 (4), 390–401.
Lakhan, S.E., Kirchgessner, A., Hofer, M., 2009. Inflammatory mechanisms in ischemic stroke: therapeutic approaches. J. Transl. Med. 7, 97.

Lapanantasin, S., Chongthammakun, S., Floyd, C.L., Berman, R.F., 2006. Effects of 17β-estradiol on intracellular calcium changes and neuronal survival after mechanical strain injury in neuronal–glial cultures. Synapse. 410 (May), 406–410.

Leon, R.L., Li, X., Huber, J.D., Rosen, C.L., 2012. Worsened outcome from middle cerebral artery occlusion in aged rats receiving 17β-estradiol. Endocrinology. 153 (7), 3386–3393.

Levin-Allerhand, J.A., Lominska, C.E., Wang, J., Smith, J.D., 2002. 17α-estradiol and 17β-estradiol treatments are effective in lowering cerebral amyloid-β levels in AβPPSWE transgenic mice. J. Alzheimers Dis. 4, 449–457.

Li, Y., Wang, S., Xia, Y., Wang, J., Pan, W., Shi, Y., et al., 2007. Neuroprotective effect of estrogen after chronic spinal cord injury in ovariectomized rats. Neural Regen. Res. 2 (8), 471–474.

Liang, K., Yang, L., Yin, C., Xiao, Z., Zhang, J., Liu, Y., et al., 2010. Estrogen stimulates degradation of beta-amyloid peptide by up-regulating neprilysin. J. Biol. Chem. 285 (2), 935–942.

Liu, X.-A., Zhu, L.-Q., Zhang, Q., Shi, H.-R., Wang, S.-H., Wang, Q., et al., 2008. Estradiol attenuates tau hyperphosphorylation induced by upregulation of protein kinase-A. Neurochem. Res. 33 (9), 1811–1820.

Lobo, R.A., 2013. Where are we 10 years after the Women's Health Initiative? J. Clin. Endocrinol. Metab. 98 (5), 1771–1780.

López Rodríguez, A.B., Mateos Vicente, B., Romero-Zerbo, S.Y., Rodriguez-Rodriguez, N., Bellini, M.J., Rodriguez de Fonseca, F., et al., 2011. Estradiol decreases cortical reactive astrogliosis after brain injury by a mechanism involving cannabinoid receptors. Cereb. Cortex. 21 (9), 2046–2055.

Manthey, D., Heck, S., Engert, S., Behl, C., 2001. Estrogen induces a rapid secretion of amyloid beta precursor protein via the mitogen-activated protein kinase pathway. Eur. J. Biochem. 268 (15), 4285–4291.

Marin, R., Guerra, B., Morales, A., Díaz, M., Alonso, R., 2003. An oestrogen membrane receptor participates in estradiol actions for the prevention of amyloid-β peptide1–40-induced toxicity in septal-derived cholinergic SN56 cells. J. Neurochem. 85 (5), 1180–1189.

Mattson, M.P., 2000. Apoptosis in neurodegenerative disorders. Nat. Rev. Mol. Cell Biol. 1 (2), 120–129.

Moore, D.W., Ashman, T.A., Cantor, J.B., Krinick, R.J., Spielman, L.A., 2010. Does gender influence cognitive outcome after traumatic brain injury? Neuropsychol. Rehabil. 20 (3), 340–354.

Nadkarni, S., McArthur, S., 2013. Oestrogen and immunomodulation: new mechanisms that impact on peripheral and central immunity. Curr. Opin. Pharmacol. 13 (4), 576–581.

Nilsen, J., Chen, S., Irwin, R.W., Iwamoto, S., Brinton, R.D., 2006. Estrogen protects neuronal cells from amyloid beta-induced apoptosis via regulation of mitochondrial proteins and function. BMC Neurosci. 7 (74), 1–14.

Nortje, J., Menon, D.K., 2004. Traumatic brain injury: physiology, mechanisms, and outcome. Curr. Opin. Neurol. 17 (6), 711–718.

Paganini-Hill, A., Henderson, V.W., 1994. Estrogen deficiency and risk of Alzheimer's disease in women. Am. J. Epidemiol. 140 (3), 256–261.

Perez, E., Liu, R., Yang, S.-H., Cai, Z.Y., Covey, D.F., Simpkins, J.W., 2005. Neuroprotective effects of an estratriene analog are estrogen receptor independent in vitro and in vivo. Brain Res. 1038 (2), 216–222.

Petanceska, S.S., Nagy, V., Frail, D., Gandy, S., 2000. Ovariectomy and 17β-estradiol modulate the levels of Alzheimer's amyloid β peptides in brain. Exp. Gerontol. 35, 1317–1325.

Piau, A., Nourhashémi, F., Hein, C., Caillaud, C., Vellas, B., 2010. Progress in the development of new drugs in Alzheimer's disease. J. Nutr. Health Aging. 15 (1), 45–57.

Prokai, L., Prokai-Tatrai, K., Perjesi, P., Zharikova, A.D., Perez, E.J., Liu, R., et al., 2003. Quinol-based cyclic antioxidant mechanism in estrogen neuroprotection. Proc. Natl. Acad. Sci. USA. 100 (20), 11741–11746.

Rao, A.K., Dietrich, A.K., Ziegler, Y.S., Nardulli, A.M., 2011. 17β-Estradiol-mediated increase in Cu/Zn superoxide dismutase expression in the brain: a mechanism to protect neurons from ischemia. J. Steroid Biochem. Mol. Biol. 127 (3–5), 382–389.

Rau, S.W., Dubal, D.B., Böttner, M., Wise, P.M., 2003. Estradiol differentially regulates c-Fos after focal cerebral ischemia. J. Neurosci. 23 (33), 10487–10494.

Raval, A.P., Saul, I., Dave, K.R., DeFazio, R.A., Perez-Pinzon, M.A., Bramlett, H., 2009. Pretreatment with a single estradiol-17β bolus activates cyclic-AMP response element binding protein and protects CA1 neurons against global cerebral ischemia. Neuroscience. 160 (2), 307–318.

Richardson, T.E., Yang, S.-H., Wen, Y., Simpkins, J.W., 2011. Estrogen protection in Friedreich's ataxia skin fibroblasts. Endocrinology. 152 (7), 2742–2749.

Ritzel, R.M., Capozzi, L.A., McCullough, L.D., 2013. Sex, stroke, and inflammation: the potential for estrogen-mediated immunoprotection in stroke. Horm. Behav. 63 (2), 238–253.

Rusa, R., Alkayed, N.J., Crain, B.J., Traystman, R.J., Kimes, A.S., London, E.D., et al., 1999. 17β-estradiol reduces stroke injury in estrogen-deficient female animals. Stroke. 30 (8), 1665–1670.

Shughrue, P.J., Merchenthaler, I., 2003. Estrogen prevents the loss of CA1 hippocampal neurons in gerbils after ischemic injury. Neuroscience. 116 (3), 851–861.

Simpkins, J.W., Rajakumar, G., Zhang, Y.Q., Simpkins, C.E., Greenwald, D., Yu, C.J., et al., 1997. Estrogens may reduce mortality and ischemic damage caused by middle cerebral artery occlusion in the female rat. J. Neurosurg. 87 (5), 724–730.

Simpkins, J.W., Perez, E., Wang, X., Yang, S., Wen, Y., Singh, M., 2009. The potential for estrogens in preventing Alzheimer's disease and vascular dementia. Ther. Adv. Neurol. Disord. 2 (1), 31–49.

Singer, C.A., Rogers, K.L., Dorsa, D.M., 1998. Modulation of Bcl-2 expression: a potential component of estrogen protection in NT2 neurons. Neuroreport. 9 (11), 2565–2568.

Singer, C.A., Figueroa-Masot, X.A., Batchelor, R.H., Dorsa, D.M., 1999. The mitogen-activated protein kinase pathway mediates estrogen neuroprotection after glutamate toxicity in primary cortical neurons. J. Neurosci. 19 (7), 2455–2463.

Soustiel, J.F., Palzur, E., Nevo, O., Thaler, I., Vlodavsky, E., 2005. Neuroprotective anti-apoptosis effect of estrogens in traumatic brain injury. J. Neurotrauma. 22 (3), 345–352.

Stoltzner, S.E., Berchtold, N.C., Cotman, C.W., Pike, C.J., 2001. Estrogen regulates bcl-x expression in rat hippocampus. Neuroreport. 12 (13), 2797–2800.

Sung, J.-H., Cho, E.-H., Min, W., Kim, M.-J., Kim, M.-O., Jung, E.-J., et al., 2010. Identification of proteins regulated by estradiol in focal cerebral ischemic injury—a proteomics approach. Neurosci. Lett. 477 (2), 66–71.

Tamatani, M., Ogawa, S., Nuñez, G., Tohyama, M., 1998. Growth factors prevent changes in Bcl-2 and Bax expression and neuronal apoptosis induced by nitric oxide. Cell Death Differ. 5 (10), 911–919.

Wang, R., Zhang, Q.-G., Han, D., Xu, J., Lü, Q., Zhang, G.-Y., 2006. Inhibition of MLK3-MKK4/7-JNK1/2 pathway by Akt1 in exogenous estrogen-induced neuroprotection against transient global cerebral ischemia by a non-genomic mechanism in male rats. J. Neurochem. 99 (6), 1543–1554.

Wen, Y., Yang, S., Liu, R., Brun-Zinkernagel, A.M., Koulen, P., Simpkins, J.W., 2004. Transient cerebral ischemia induces aberrant neuronal cell cycle re-entry and Alzheimer's disease-like tauopathy in female rats. J. Biol. Chem. 279 (21), 22684–22692.

Xu, H., Gouras, G.K., Greenfield, J.P., Vincent, B., Naslund, J., Mazzarelli, L., et al., 1998. Estrogen reduces neuronal generation of Alzheimer β-amyloid peptides. Nat. Med. 4 (4), 447–451.

Xu, H., Wang, R., Zhang, Y.-W., Zhang, X., 2006. Estrogen, beta-amyloid metabolism/trafficking, and Alzheimer's disease. Ann. N. Y. Acad. Sci. 1089, 324–342.

Yang, S.H., Shi, J., Day, A.L., Simpkins, J.W., 2000. Estradiol exerts neuroprotective effects when administered after ischemic insult. Stroke. 31 (3), 745–749, discussion 749–750.

Yeung, J.H.H., Mikocka-Walus, A.A., Cameron, P.A., Poon, Q.S., Ho, H.F., Chang, A., et al., 2011. Protection from traumatic brain injury in hormonally active women vs. men of a similar age. Arch. Surg. 146 (4), 436–442.

Zhang, Q.-G., Wang, R., Khan, M., Mahesh, V., Brann, D.W., 2008. Role of Dickkopf-1, an antagonist of the Wnt/beta-catenin signaling pathway, in estrogen-induced neuroprotection and attenuation of tau phosphorylation. J. Neurosci. 28 (34), 8430–8441.

Zhang, Y.Q., Shi, J., Rajakumar, G., Day, A.L., Simpkins, J.W., 1998. Effects of gender and estradiol treatment on focal brain ischemia. Brain Res. 784 (1–2), 321–324.

Zhang, Z., Simpkins, J.W., 2010. Okadaic acid induces tau phosphorylation in SH-SY5Y cells in an estrogen-preventable manner. Brain Res. 1345, 176–181.

Zhao, L., Yao, J., Mao, Z., Chen, S., Wang, Y., Brinton, R.D., 2011. 17β-Estradiol regulates insulin-degrading enzyme expression via an ERβ/PI3-K pathway in hippocampus: relevance to Alzheimer's prevention. Neurobiol. Aging. 32 (11), 1949–1963.

Zheng, H., Xu, H., Uljon, S.N., Gross, R., Hardy, K., Gaynor, J., et al., 2002. Modulation of A(beta) peptides by estrogen in mouse models. J. Neurochem. 80 (1), 191–196.

Zlotnik, A., Leibowitz, A., Gurevich, B., Ohayon, S., Boyko, M., Klein, M., et al., 2012. Effect of estrogens on blood glutamate levels in relation to neurological outcome after TBI in male rats. Intensive Care Med. 38 (1), 137–144.

CHAPTER

8

Neuroprotection with Estradiol in Experimental Perinatal Asphyxia: A New Approach

George Barreto[1], *Ezequiel Saraceno*[2], *Janneth Gonzalez*[1], *Rodolfo Kolliker*[2], *Rocío Castilla*[2], *and Francisco Capani*[2,3,4]

[1]Departamento de Nutrición y Bioquímica, Facultad de Ciencias, Pontificia Universidad Javeriana, Bogotá, Colombia [2]Laboratorio de Citoarquitectura e Injuria Neuronal, Instituto de Investigaciones Cardiológicas "Prof. Dr. Alberto C. Taquini" (ININCA), UBA-CONICET, Buenos Aires, Argentina [3]Departamento de Biología, Universidad Argentina John F Kennedy, Buenos Aires, Argentina [4]Facultad de Psicología, Universidad Católica Argentina, Buenos Aires, Argentina

PERINATAL ASPHYXIA OVERVIEW

Perinatal asphyxia (PA) brain-induced injury is one of the most frequent causes of morbidity and mortality in term and preterm neonates, accounting for 23% of neonatal deaths globally (Lawn et al., 2005). Following PA, approximately 45% of newborns die and 25% have permanent neurological deficits including cerebral palsy, mental retardation and developmental delay, learning disabilities, eyesight and hearing problems, and different issues in school readiness (Amiel-Tison and Ellison, 2010).

Brain injury occurring early during development results in significant damage in different areas of the central nervous systems (CNS), such as: cortex, hippocampus, neostriatum, cerebellum and substantia nigra (Capani et al., 2009). The type and distribution of human brain lesions

K.A. Duncan (Ed): *Estrogen Effects on Traumatic Brain Injury*.
DOI: http://dx.doi.org/10.1016/B978-0-12-801479-0.00008-5

differ markedly between premature and term babies, likely as a consequence of the stage of brain maturation and subsequent regional vulnerability, as described in different previous studies (Miller and Ferriero, 2009; Verger et al., 2001; Yager and Thornhill, 1997).

In the last few years different mechanisms of injury have been proposed for the immature versus the adult brain. The most obvious difference is that apoptotic mechanisms are several-fold more pronounced in immature animals (Li et al., 2011). A substantial body of evidence suggests that the developing brain shows marked susceptibility to both oxidative stress and neuronal apoptosis, which may be related to this age-dependent injury vulnerability (Bayir et al., 2006; Blomgren et al., 2003; Blomgren et al., 2007; Ikonomidou and Kaindl, 2011; Li et al., 2010; Zhu et al., 2005).

The mechanisms that cause neurological damage after PA are divided schematically into three metabolic phases (Gunn et al., 1997; Potts et al., 2006; Roelfsema et al., 2004). Hypoxia leads to primary energy failure (phase 1). Shortly after reoxygenation, aerobic metabolism is re-established (phase 2). However, as a result of a cascade of cellular mechanisms (Hobbs et al., 2008; Kittaka et al., 1997), after 6–24 hours mitochondrial energy production begins to fail one more time. This secondary energy failure (phase 3) is induced for 24–48 hours after the hypoxic event. The damage that occurs during phase 3 is considerable and leads to deep cell damage (Vannucci et al., 2004).

All the modifications induced by PA are related to biological reactions that result in partial recovery, but also in an overexpression of several metabolic, molecular and cell cascades, extending the energy deficit and oxidative stress, associated with further neuronal damage, apoptosis (with several genes such as Bcl2, Bax, among others) or necrosis. Molecular and cell cascades involved in the removal of damaged cells are ubiquitination, peroxisomal, and caspase pathways, whereas some compensatory cascades include multiple mechanisms of DNA repair, for prevention of loss or salvage of cells. On the other hand, oxidative stress is strongly connected with reoxygenation after PA, resulting in both the overactivation and the inactivation of a number of buffering enzymes, including those modulating the activity of mitochondria (Davis et al., 2004; Gitto et al., 2002). In addition, in the clinical situation, resuscitation may even imply hyperoxemia, leading to a further production of free radicals and oxidative stress, worsening brain injury (Davis et al., 2004; Gitto et al., 2002).

In the last few years, we have used a model that induces global severe PA. The advantages of this model are threefold: first, it mimics asphyxia just in the moment of delivery; second, it allows for the study of both short- and long-term effects, since it is a noninvasive procedure; third, it is well reproducible across laboratories. The most obvious and serious short-term consequence of PA observed in this model is mortality. At 37°C, an asphyctic period longer than 20 minutes and the consequent ATP deficit lead to the activation of anaerobic glycolysis and the accumulation of acidosis in the extracellular compartments (Capani et al., 2001). Prolonged PA leads to an increase in hypoxia inducible factor (HIF-1) expression and a decrease in transcription and translation (Capani et al., 2003). Reoxygenation is associated with death, probably induced by glutamate overactivation and excessive free radical release (Capani et al., 2001; Capani et al., 2003). Rats subjected to asphyxia for 20 min at 37°C and monitored by microdialysis showed chronic defects in neurotransmitters, such as a decrease in dopamine, and aspartate and glutamate release (Capani et al., 2003). We also found an increase in NO in neostriatum and neocortex in the short and long term (Capani et al., 1997). In addition, we described increased ubiquitination levels and neurodegeneration 6 months after PA in neostriatum (Capani et al., 2009). Alterations in cytoskeletal organization were observed 4 months after PA (Saraceno et al., 2010). Moreover, behavioral deficits such as exploration of new environments, spatial reference and working memory were observed 3 months after PA (Galeano et al., 2011). Finally, we showed that the actin cytoskeleton is highly modified 30 days after PA (Capani et al., 2008; Saraceno et al., 2012). Thus, we think synaptic actin could trigger long-term changes induced by PA. Figure 8.1 summarizes the most important mechanisms inherent to the pathophysiology of PA.

NEUROPROTECTIVE ROLE OF ESTRADIOL IN RAT MODELS OF HYPOXIA

Although different treatments have been used so far, only hypothermia is recommended as a routine in clinical practice (Hagmann et al., 2011). In the last few years, the ovarian hormone 17β-estradiol has emerged as a potential neuroprotector agent against several neuropathological diseases including Parkinson's, Alzheimer's, multiple sclerosis and stroke (Bourque et al., 2009; Garcia-Segura and Balthazart, 2009; Kipp and Beyer, 2009; Pike et al., 2009; Suzuki et al., 2009). In recent work we have shown the repair of chronic neurodegenerative modifications in 4-month-old male Sprague-Dawley rats submitted to PA and injected for 3 consecutive days with 17β-estradiol or vehicle (Saraceno et al., 2010). As described before, we observed, in the hippocampus of PA and vehicle-treated animals, classical astrogliosis, focal swelling and fragmented appearance of MAP-2 immunoreactive dendrites, decreased MAP-2 immunoreactivity and decreased phosphorylation of high and medium molecular weight neurofilaments, as compared to control

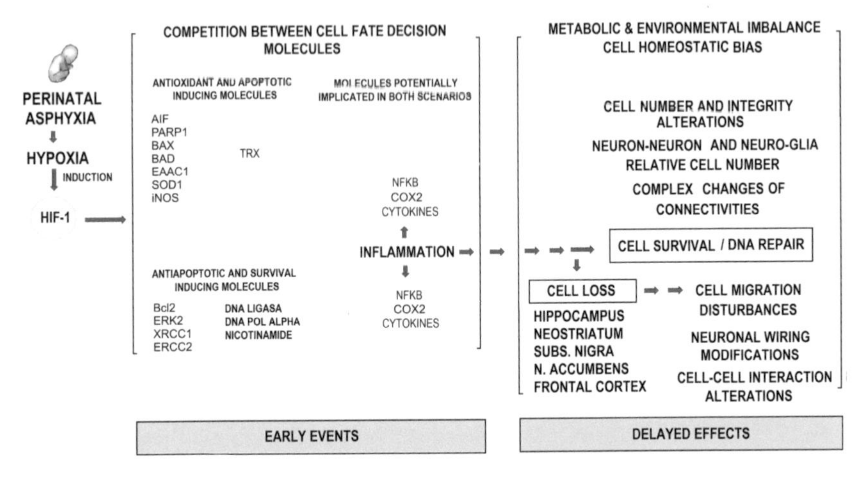

FIGURE 8.1 **Schematic molecular and cell events elicited by perinatal asphyxia.** Perinatal asphyxia (PA) decreases oxygen saturation in blood, leading to a switch from aerobic to a less efficient anaerobic metabolism involving lactate accumulation, acidosis and cell death. PA associated hypoxia impairs gene expression, decreases transcription and translation, as well as activation of genes, such as hypoxia inducible factor (HIF) and its target molecules. Two main molecular and cell cascades are elicited by perinatal hypoxic insults. One of them leads to removal of cells damaged by their inadequate supply of oxygen, triggering activation of ubiquitination, peroxisomal, and caspase pathways producing cell loss through apoptosis or necrosis, the latter encompassing pro-inflammatory. The other cascade is activated to preserve homeostasis reducing cell loss by multiple mechanisms of DNA repair. Various mechanisms of action could support these compensatory mechanisms. In any case, both cell loss and cell rescue eventually entail more or less subtle and complex consequences on brain development, neuronal wiring, and neuron glia interactions. These consequences seem to be reinforced by additional negative impact during development (puberty and adolescence), resulting in psychiatric disorders if not aggravating neurological deficits. AIF: Apoptosis inducing factor (caspase independent).

animals. These alterations are well-known markers of neurodegenerative damage. We observed them either just after the end of the treatment or until one month of treatment had passed. On the other hand, treatment with estradiol reversed these alterations (Figure 8.2). These properties of estradiol have been described previously when the hormone was administered before or shortly after the induction of brain damage (Barreto et al., 2009; Shepherd et al., 2002).

We showed that the late administration of estradiol may reduce, in adult animals, chronic neurodegenerative modifications induced by PA in the early stages of development. Mechanisms involved in the neuroprotective properties of estradiol are not fully understood yet. Since NF proteins are preferential targets for oxidative stress (Hand and Rouleau, 2002), estrogens can induce their neuroprotective function modulating free radical production. In addition, a substantial body of evidence

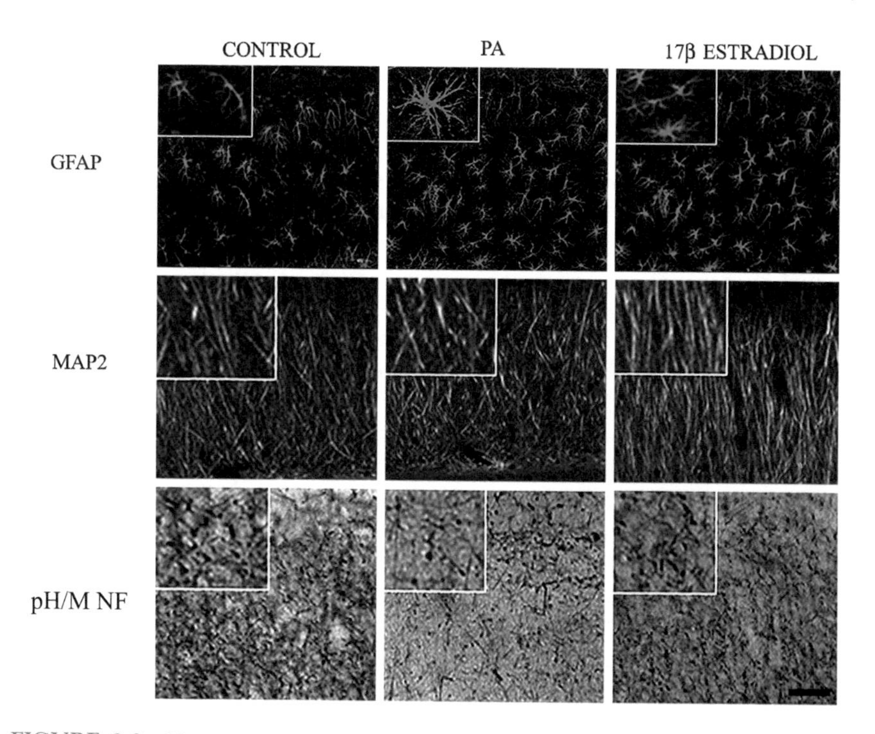

FIGURE 8.2 **Neuroprotective effect of 17β estradiol over the cytoskeletal modifications induced by PA in hippocampus.** Three different markers for astrocytes (GFAP), dendritic microtubules (MAP-2) and axonal phosphorylated high and medium molecular weight neurofilaments. Asphyctic animals injected with 17β estradiol showed similar morphological organization than the control, suggesting that this treatment might revert neurodegenerative alterations induced by PA. Scale bar, 10 μm.

suggests that estrogens are related to the facilitation and preservation of neurite growth during the development of the CNS (Dominguez et al., 2004). Finally, 17β-estradiol effects may be mediated by ERK activation, which is involved in the maintenance of dendritic tree organization and neuronal morphology in pro-apoptotic conditions (Miñano et al., 2008).

SIGNALING PATHWAYS OF THE ESTRADIOL NEUROPROTECTION

As mentioned above, several clinical studies suggest that estradiol acts as a potent growth and protective factor in the adult brain. Estrogen protects neurons against numerous traumatic or chronic neurological and mental diseases in humans (Saunders-Pullman et al., 1999; Waring et al., 1999).

Estrogen increases the viability, survival and differentiation of primary neuronal cultures deprived of growth factors and protects neuronal cell lines and primary neuronal cultures from anoxia (Garcia-Segura et al., 2001).

The biological effects of estrogens are exerted by different kinds of mechanisms. Estradiol (at supraphysiological concentrations) has antioxidant properties that depend on the presence of the hydroxyl group in the C3 position on the A ring of the steroid molecule and are independent of the activation of estrogen receptors (ERs) (Behl and Holsboer, 1996). On the other hand, estradiol binds to ERs and can additionally operate by two different types of mechanism. In the classical one, estrogen binds to ERα or ERβ, both members of a large superfamily of nuclear receptors which act as ligand-activated transcription factors. After estrogen binding, they dimerize and bind to specific response elements (EREs) located in the target gene promoters (Nilsson et al., 2001). In addition, ERs can regulate (activating or repressing) gene expression by modulating the function of other classes of transcription factors through protein–protein interactions in the nucleus (Gottlicher et al., 1998) regulating the expression of a large number of estrogen-responsive genes that do not contain EREs, without directly binding to DNA.

Furthermore, a number of estrogen effects are so rapid that they cannot depend on the activation of RNA and protein synthesis. These actions are known as nongenomic and are mediated by membrane-associated ERs and associated with the activation of various protein-kinase cascades (Losel and Wehling, 2003). However, nongenomic actions may indirectly influence gene expression, through the activation of signal transduction pathways that eventually act (through protein kinase-mediated phosphorylation) on target transcription factors. Thus, the possible convergence of genomic and nongenomic actions at multiple response elements of gene promoters provides an extremely fine degree of control in the regulation of transcription by ERs.

Some researchers have suggested that the nongenomic actions of estrogens are mediated by a subpopulation of the classical ERα and ERβ, located at the plasma membrane (Pappas et al., 1995; Razandi et al., 1999) or by a member if the 7-transmembrane G-protein-coupled receptor family, GPR30 (Filardo and Thomas, 2005), among others (Toran-Allerand et al., 2002).

The molecular mechanisms underlying the nongenomic actions of estrogens are specific for each cell type and include the mobilization of intracellular calcium (Improta-Brears et al., 1999), and the stimulation of adenylate cyclase activity and cAMP production

(Aronica et al., 1994; Razandi et al., 1999). The activation of MAPK signaling by 17β-estradiol has been extensively studied in several cell types, including neuroblastoma cells (Watters et al., 1997). 17β-estradiol also activates the phosphoinositol (PI) 3-kinase signaling pathway in neuron cells (Varea et al., 2009).

ERs at the plasma membrane associate with the scaffold protein caveolin-1 (Chambliss et al., 2000; Razandi et al., 2002), and with a variety of proximal signaling molecules such as G proteins (Razandi et al., 1999; Wyckoff et al., 2001), Src kinase, and *ras* (Migliaccio et al., 1998), the p85α regulatory subunit of PI3-kinase (Simoncini et al., 2000) and Shc (Song et al., 2002). Furthermore, the scaffold protein MNAR promotes the interaction of 17β-estradiol-activated ERα with Src kinase, leading to an increase in Src kinase activity, and consequently the activation of the MAPK signaling pathway (Wong et al., 2002).

Membrane ERs also activate membrane tyrosine kinase receptors in various cell types. ERα activated by 17β-estradiol interacts directly with the IGF-I receptor, leading to its activation, and hence the activation of the MAPK signaling pathway (Kahlert et al., 2000). ERα also interacts directly with ErbB2 (HER-2/neu) (Chung et al., 2002). In addition, 17β-estradiol-bound ERα activates the epidermal growth factor (EGF) receptor by a mechanism that involves the activation of G proteins, Src kinase, and matrix metalloproteinases, leading to an increase in MAPK and Akt activities (Razandi et al., 2003).

Estrogens may use different signaling pathways depending on the cellular context, and in this way evoke distinct gene responses in different types of target cells. In the same line, depending on the type of neurodegenerative insult, the brain area involved, the time after injury, and the sex and age of the individual, the neuroprotective effects of estradiol *in vivo* may involve different receptor subtypes as well as mechanisms independent of ERs.

It has been demonstrated that estradiol decreases the response to various forms of insult and that the brain itself upregulates both estrogen synthesis and estrogen receptor expression at injury sites, which suggests that these receptors are involved in the response of neural tissue to injury and may, therefore, mediate neuroprotective effects of estradiol *in vivo*.

We studied the signaling pathway used by estradiol in the murine model of PA described here previously. To that end, we first evaluated the effect of estradiol on cell viability in a model of hypoxia *in vitro*. We used human SH-SY5Y neuroblastoma cells subjected to hypoxia by CoCl2 (250 μg/mL), a hypoxic mimetic agent, and then treated with 17β-estradiol (250 nM). An increment in cell survival was determined in estrogen-treated cells (unpublished results). With the aim of determining the estrogen receptor involved in estradiol neuroprotective action, we treated the hypoxic SH-SY5Y cells with 17β-estradiol in the presence of an inhibitor of both ERα and ERβ (ICI182780, 10 nM) which also acts as an agonist of GPR30 receptor. The increase in cell survival promoted by estradiol was prevented by the presence of ICI182780, indicating that ERα and/or ERβ are involved in its mechanism of action. To discriminate which of the classical ERs participates in the neuroprotective activity, the hypoxic cells were treated with ERα or ERβ agonists (PPT, 10 nM or DPN, 10 nM, respectively) instead of 17β-estradiol. The neuroprotection by estradiol was mimicked by PPT, which indicates that its action is mediated by ERα.

Taking into account the results obtained using SH-SY5Y cells, the hippocampal expression of ERs was analyzed in AP and 17β-estradiol-treated rats. No changes in ERα, ERβ or GPR30 were detected in animals submitted to PA, but an increase in ERα was observed when PA rats were treated with 17β-estradiol.

Since the interaction between ERα and IGF-IR is well demonstrated in the promotion of neuronal survival and in the response of neural tissue to injury, we studied the participation of this pathway in the repair mechanism of the morphological alterations induced by PA. IGF-IR expression was increased when PA rats were treated with 17β-estradiol, which further demonstrates the participation of this pathway in estradiol neuroprotective actions. In addition, we demonstrated that the PI3K/Akt/GSK3 signaling pathway is activated, as an increase in Akt and GSK3 phosphorylation, and in β-catenin protein expression was observed by late estradiol treatment in PA animals. Thus, the neuroprotective actions of estradiol in PA animals are produced by the activation of ERβ and the IGF-IR pathway.

CONCLUSION

Estradiol treatment not only prevents neuronal damage, but also may limit the neurodegenerative modifications induced by PA in the early stage of development. Because estradiol may have numerous undesirable peripheral effects, further studies should address the effect of selective estrogen receptor modulators or nonfeminizing estrogens, which may represent a more adequate therapeutic approach. In addition, new knowledge about cellular repair mechanisms can also pave the way for types of treatment that do not merely limit damage, but can also repair defects in the immature nervous system.

References

Amiel-Tison, C., Ellison, P., 2010. Birth asphyxia in the fullterm newborn: early assessment and outcome. Dev. Med. Child Neurol. 28, 671–682.

Aronica, S.M., Kraus, W.L., Katzenellenbogen, B.S., 1994. Estrogen action via the cAMP signalling pathway: stimulation of adenylate cyclase and cAMP-regulated gene transcription. Proc. Natl. Acad. Sci. USA 91, 8517–8521.

Barreto, G., Santos-Galindo, M., Diz-Chaves, Y., Pernía, O., Carrero, P., Azcoitia, I., et al., 2009. Selective estrogen receptor modulators decrease reactive astrogliosis in the injured brain: effects of aging and prolonged depletion of ovarian hormones. Endocrinology. 150, 5010–5015.

Bayir, H., Kochanek, P.M., Kagan, V.E., 2006. Oxidative stress in immature brain after traumatic brain injury. Dev. Neurosci. 28, 420–431.

Behl, C.F., Holsboer, F., 1996. The female sex hormone oestrogen as a neuroprotectant. Trends Pharmacol. Sci. 20, 441–444.

Blomgren, K., Zhu, C., Hallin, U., Hagberg, H., 2003. Mitochondria and ischemic reperfusion damage in the adult and in the developing brain. Biochem. Biophys. Res. Commun. 304, 551–559.

Blomgren, K., Leist, M., Groc, L., 2007. Pathological apoptosis in the developing brain. Apoptosis. 12, 993–1010.

Bourque, M., Dluzen, D.E., Di Paolo, T., 2009. Neuroprotective actions of sex steroids in Parkinson's disease. Front. Neuroendocrinol. 30, 142–157.

Capani, F., Loidl, F., López-Costa, J.J., Selvin-Testa, A., Saavedra, J.P., 1997. Ultrastructural changes in nitric oxide synthase immunoreactivity in the brain of rats subjected to perinatal asphyxia: neuroprotective effects of cold treatment. Brain Res. 775, 11–23.

Capani, F., Loidl, C.F., Aguirre, F., Piehl, L., Facorro, G., Hager, A., et al., 2001. Changes in reactive oxygen species (ROS) production in rat brain during global perinatal asphyxia: an ESR study. Brain Res. 914, 204–207.

Capani, F., Loidl, C.F., Piehl, L.L., Facorro, G., De Paoli, T., Hager, A., 2003. Long term production of reactive oxygen species during perinatal asphyxia in the rat central nervous system: effects of hypothermia. Int. J. Neurosci. 113, 641–654.

Capani, F., Saraceno, G.E., Boti, V.R., Aón-Bertolino, L., Fernández, J.C., Gato, F., et al., 2008. A tridimensional view of the organization of actin filaments in the central nervous system by use of fluorescent photooxidation. Biocell. 32, 1–14.

Capani, F., Saraceno, G.E., Botti, V., Aón-Bertolino, L., de Oliveira, D.M., Barreto, G., et al., 2009. Protein ubiquitination in postsynaptic densities after hypoxia in rat neostriatum is blocked by hypothermia. Exp. Neurol. 219, 404–413.

Chambliss, K.L., Yuhanna, I.S., Mineo, C., Liu, P., German, Z., Sherman, T.S., et al., 2000. Estrogen receptor α and endothelial nitric oxide synthase are organized into a functional signalling module in caveolae. Circ. Res. 87, E44–E52.

Chung, Y.L., Sheu, M.L., Yang, S.C., Lin, C.H., Yen, S.H., 2002. Resistance to tamoxifen-induced apoptosis is associated with direct interaction between Her2/neu and cell membrane estrogen receptor in breast cancer. Int. J. Cancer. 97, 306–312.

Davis, P.G., Tan, A., O'Donnell, C.P., Schulze, A., 2004. Resuscitation of new-born infants with 100% oxygen or air: asystematic review and meta-analysis. Lancet. 364, 1329–1333.

Dominguez, R., Jalali, C., de Lacalle, S., 2004. Morphological effects of estrogen on cholinergic neurons in vitro involves activation of extracellular signal-regulated kinases. J. Neurosci. 24, 982–990.

Filardo, E.J., Thomas, P., 2005. GPR30: a seven-membrane-spanning estrogen receptor that triggers EGF release. Trends Endocrinol. Metab. 16, 362–367.

Galeano, P., Blanco Calvo, E., Madureira de Oliveira, D., Cuenta, L., Kamenetzky, G.V., Mustaca, A.E., et al., 2011. Long-lasting effects of perinatal asphyxia on exploration, memory and incentive downshift. Int. J. Dev. Neurosci. 29 (6), 609–619.

Garcia-Segura, L.M., Balthazart, J., 2009. Steroids and neuroprotection: new advances. Front. Neuroendocrinol. 30, v–ix.

Garcia-Segura, L.M., Azcoitia, I., DonCarlos, L.L., 2001. Neuroprotection by estradiol. Prog. Neurobiol. 63, 29–60.

Gitto, E., Reiter, R.J., Karbownik, M., Tan, D.-X., Gitto, P., Barberi, S., et al., 2002. Causes of oxidative stress in the pre- and perinatal period. Biol. Neonate. 81, 146–157.

Gottlicher, M., Heck, S., Herrlich, P., 1998. Transcriptional cross-talk, the second mode of steroid hormone receptor action. J. Mol. Med. 76, 480–489.

Gunn, A.J., Gunn, T.R., de Haan, H.H., Van Reempts, J.L., Van Belle, H., Hasaart, T.H., 1997. Dramatic neuronal rescue with prolonged selective head cooling after ischemia in fetal lambs. J. Clin. Invest. 99, 248–256.

Hagmann, C.F., Brotschi, B., Bernet, V., Berger, T.M., Robertson, N.J., 2011. Hypothermia for perinatal encephalopathy. Swiss Med. Wkly. 141, w13145.

Hand, C.K., Rouleau, G.A., 2002. Familial amyotrophic lateral sclerosis. Muscle Nerve. 25, 35–139.

Hobbs, C., Thoresen, M., Tucker, A., Aquilina, K., Chakkarapani, E., Dingley, J., 2008. Xenon hypothermia combine additively, offering long-term functional and histopathologic neuroprotection after neonatal hypoxia/ischemia. Stroke. 39, 1307–1313.

Ikonomidou, C., Kaindl, A.M., 2011. Neuronal death and oxidative stress in the developing brain. Antioxid. Redox Signal. 14, 1535–1550.

Improta-Brears, T., Whorton, A.R., Codazzi, F., York, J.D., Meyer, T., McDonnell, D.P., 1999. Estrogen-induced activation of mitogen-activated protein kinase requires mobilization of intracellular calcium. Proc. Natl. Acad. Sci. USA 96, 4686–4691.

Kahlert, S., Nuedling, S., van Eickels, M., Vetter, H., Meyer, R., Grohe, C., 2000. Estrogen receptor α rapidly activates the IGF-1 receptor pathway. J. Biol. Chem. 275, 18447–18453.

Kipp, M., Beyer, C., 2009. Impact of sex steroids on neuroinflammatory processes and experimental multiple sclerosis. Front. Neuroendocrinol. 30, 188–200.

Kittaka, M., Giannotta, S.L., Zelman, V., Correale, J.D., DeGiorgio, C.M., Weiss, M.H., et al., 1997. Attenuation of brain injury and reduction of neuron-specific enolase by nicardipine in systemic circulation following focal ischemia and reperfusion in a rat model. J. Neurosurg. 87, 731–737.

Lawn, J.E., Cousens, S., Zupan, J., Team, L.N.S.S., 2005. 4 million neonatal deaths: When? Where? Why? Lancet. 365, 891–900.

Li, H., Li, Q., Du, X., Sun, Y., Wang, X., Kroemer, G., et al., 2011. Lithium-mediated long-term neuroprotection in neonatal rat hypoxia-ischemia is associated with anti-inflammatory effects and enhanced proliferation and survival of neural stem/progenitor cells. JCBFM. 31, 2106–2115.

Li, Q., Li, H., Roughton, K., Wang, X., Kroemer, G., Blomgren, K., et al., 2010. Lithium reduces apoptosis and autophagy after neonatal hypoxia-ischemia. Cell Death Dis. 1, e56.

Losel, R., Wehling, M., 2003. Nongenomic actions of steroid hormones. Nat. Rev. Mol. Cell Biol. 4, 46–56.

Migliaccio, A., Piccolo, D., Castoria, G., Di Domenico, M., Bilanciom, A., Lombardi, M., et al., 1998. Activation of the Src/p21ras/Erk pathway by progesterone receptor via cross-talk with estrogen receptor. EMBO J. 17, 2008–2018.

Miller, S.P., Ferriero, D.M., 2009. From selective vulnerability to connectivity: insights from newborn brain imaging. Trends Neurosci. 32, 496–505.

Miñano, A., Xifró, X., Pérez, V., Barneda-Zahonero, B., Saura, C.A., Rodríguez-Alvarez, J., 2008. Estradiol facilitates neurite maintenance by a Src/Ras/ERK signalling pathway. Mol. Cell. Neurosci. 39, 143–151.

Nilsson, S., Makela, S., Treuter, E., Tujague, M., Thomsen, J., Andersson, G., et al., 2001. Mechanisms of estrogen action. Physiol. Rev. 81, 1535–1565.

Pappas, T.C., Gametchu, B., Watson, C.S., 1995. Membrane estrogen receptors identified by multiple antibody labelling and impeded-ligand binding. FASEB J. 9, 404–410.

Pike, C.J., Carroll, J.C., Rosario, E.R., Barron, A.M., 2009. Protective actions of sex steroid hormones in Alzheimer's disease. Front. Neuroendocrinol. 30, 239–258.

Potts, M., Koh, S.-E., Whetstone, W., Walker, B., Yoneyama, T., Claus, C.P., et al., 2006. Traumatic injury to the immature brain: inflammation, oxidative injury, and iron-mediated damage as potential therapeutic targets. Neuroreport. 3, 143–153.

Razandi, M., Pedram, A., Greene, G.L., Levin, E.R., 1999. Cell membrane and nuclear estrogen receptors (ERs) originate from a single transcript: studies of ERα and ERβ expressed in Chinese hamster ovary cells. Mol. Endocrinol. 13, 307–319.

Razandi, M., Oh, P., Pedram, A., Schnitzer, J., Levin, E.R., 2002. ERs associate with and regulate the production of caveolin: implications for signalling and cellular actions. Mol. Endocrinol. 16, 100–115.

Razandi, M., Pedram, A., Park, S.T., Levin, E.R., Oh., P., Schnitzer, J., 2003. Proximal events in signalling by plasma membrane estrogen receptors ERs associate with and regulate the production of caveolin: implications for signalling and cellular actions. J. Biol. Chem. 278, 2701–2712.

Roelfsema, V., Bennet, L., George, S., Wu, D., Guan, J., Veerman, M., et al., 2004. Window of opportunity of cerebral hypothermia for postischemic white matter injury in the near-term fetal sheep. J. Cereb. Blood Flow Metab. 24, 877–886.

Saraceno, G., Ayala, M., Badorrey, M., Holubiec, M., Romero, J., Galeano, P., et al., 2012. Effects of perinatal asphyxia on rat striatal cytoskeleton. Synapse. 66 (1), 9–19.

Saraceno, G.E., Bertolino, M.L., Galeano, P., Romero, J.I., Garcia-Segura, L.M., Capani, F., 2010. Estradiol therapy in adulthood reverses glial and neuronal alterations caused by perinatal asphyxia. Exp. Neurol. 223, 615–622.

Saunders-Pullman, R., Gordon-Elliott, J., Parides, M., Fahn, S., Saunders, H.R., Bressman, S., 1999. The effect of estrogen replacement on early Parkinson's disease. Neurology. 52, 1417–1421.

Shepherd, C.E., McCann, H., Thiel, E., Halliday, G.M., 2002. Neurofilament-immunoreactive neurons in Alzheimer's disease and dementia with Lewy bodies. Neurobiol. Dis. 9, 249–257.

Simoncini, T., Hafezi-Moghadam, A., Brazil, D.P., Ley, K., Chin, W.W., Liao, J.K., 2000. Interaction of oestrogen receptor with the regulatory subunit of phosphatidylinositol-3-OH kinase. Nature. 407, 538–541.

Song, R.X., McPherson, R.A., Adam, L., Bao, Y., Shupnik, M., Kumar, R., et al., 2002. Linkage of rapid estrogen action to MAPK activation by ERα-Shc association and Shc pathway activation. Mol. Endocrinol. 16, 116–127.

Suzuki, S., Brown, C.M., Wise, P.M., 2009. Neuroprotective effects of estrogens following ischemic stroke. Front. Neuroendocrinol. 30, 201–211.

Toran-Allerand, C.D., Guan, X., MacLusky, N.J., Horvath, T.L., Diano, S., Singh, M., et al., 2002. ER-X: a novel, plasma membrane-associated, putative estrogen receptor that is regulated during development and after ischemic brain injury. J. Neurosci. 22, 8391–8401.

Vannucci, R.C., Towfighi, J., Vannucci, S.J., 2004. Secondary energy failure after cerebral hypoxia-ischemia in the immature rat. J. Cereb. Blood Flow Metab. 24, 1090–1097.

Varea, O., Garrido, J.J., Dopazo, A., Mendez, P., Garcia-Segura, L.M., Wandosell, F., 2009. PLoS One. 4 (4), e5153.

Verger, K., Junqué, C., Levin, H.S., Jurado, M.A., Pérez-Gómez, M., Bartrés-Faz, D., et al., 2001. Correlation of atrophy measures on MRI with neuropsychological sequelae in children and adolescents with traumatic brain injury. Brain Inj. 15, 211–221.

Waring, S.C., Rocca, W.A., Petersen, R.C., O'Brien, P.C., Tangalos, E.G., Kokmen., E., 1999. Postmenopausal estrogen replacement therapy and risk of AD: a population-based study. Neurology. 52, 965–970.

Watters, J.J., Campbell, J.S., Cunningham, M.J., Krebs, E.G., Dorsa, D.M., 1997. Rapid membrane effects of steroids in neuroblastoma cells: effects of estrogen on mitogen activated protein kinase signalling cascade and c-fos immediate early gene transcription. Endocrinology. 138, 4030–4033.

Wong, C.W., McNally, C., Nickbarg, E., Komm, B.S., Cheskis, B.J., 2002. Estrogen receptor-interacting protein that modulates its nongenomic activity-crosstalk with Src/Erk phosphorylation cascade. Proc. Natl. Acad. Sci. USA 99, 14783–14788.

Wyckoff, M.H., Chambliss, K.L., Mineo, C., Yuhanna, I.S., Mendelsohn, M.E., Mumby, S.M., et al., 2001. Plasma membrane estrogen receptors are coupled to endothelial nitric-oxide synthase through Gα (i). J. Biol. Chem. 276, 27071–27076.

Yager, J.Y., Thornhill, J.A., 1997. The effect of age on susceptibility to hypoxic-ischemic brain damage. Neurosci. Biobehav. Rev. 21, 167–174.

Zhu, C., Wang, X., Xu, F., Bahr, B.A., Shibata, M., Uchiyama, Y., et al., 2005. The influence of age on apoptotic and other mechanisms of cell death after cerebral hypoxia-ischemia. Cell Death Differ. 12, 162–176.

CHAPTER

9

Cerebrovascular Stroke: Sex Differences and the Impact of Estrogens

Farida Sohrabji

Women's Health in Neuroscience Program, Neuroscience and Experimental Therapeutics, Texas A&M Health Science Center, College of Medicine, Bryan, Texas, USA

WHAT IS STROKE?

Stroke occurs when blood supply to the brain is interrupted, resulting in the rapid death of neurons, which causes a broad range of neurologic problems including loss of sensory-motor function, depression, dementia, epilepsy and even death. Stroke can be classified broadly into two types: ischemic stroke and hemorrhagic stroke (see Table 9.1). In an ischemic stroke, a portion of the brain is deprived of blood (and therefore glucose and oxygen). An ischemic stroke typically occurs when a clot obstructs blood flow in a blood vessel and is by far the most common type of stroke (87%). The second type is a hemorrhagic stroke, where a blood vessel ruptures, causing blood to flood into the brain, where it eventually clots. Hemorrhagic strokes are more rare (13%) but they are also more severe and more likely to result in death.

ROLE OF ESTROGEN

Estrogen is a steroid hormone, principally produced by the ovaries in females. Estrogen is derived from the aromatization of testosterone; thus cells that produce aromatase can locally produce estradiol as well.

K.A. Duncan (Ed): Estrogen Effects on Traumatic Brain Injury.
DOI: http://dx.doi.org/10.1016/B978-0-12-801479-0.00009-7

TABLE 9.1 Clinical and Experimental Findings in Stroke Regarding Sex Differences

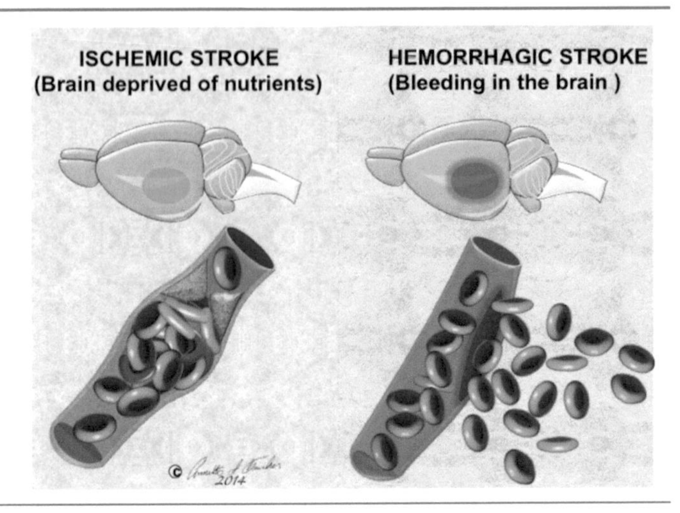

	Clinical findings	
Incidence	**87%**	**13%**
Sex difference in incidence	Below 69 years: Occurs more often in males than females After 85 years: Occurs more often in females than males	SAH: Higher relative risk in females ICH: Similar rates in males and females
Severity	Aging females have worse outcomes than aging males	Females have greater mortality than males
Effect of estrogen	Increases the incidence in menopausal women	Does not increase the incidence in menopausal women
	Experimental findings	
Sex differences	Young females have smaller infarcts than young males Aging females have greater mortality than aging males	Males have greater mortality, greater bleeding than females
Effect of estrogen	Improves outcomes in young males and females Contradictory evidence for aging females	Improves outcomes in males and females

In the central nervous system, astrocytes upregulate the expression of aromatase after stroke (Carswell et al., 2005), and have been shown to synthesize estradiol (Azcoitia et al., 2010).

In females, estrogen synthesis and release from the ovaries occurs after puberty (menarche), and, throughout adulthood, ovarian hormone levels rise and fall in conjunction with the menstrual cycle. In the follicular phase, estrogen levels rise, while progesterone levels rise during the luteal phase of the menstrual cycle. Thus, females are exposed to cyclic levels of estrogen throughout their reproductive years. At menopause, which occurs on average at 51 years of age, estrogen levels decrease significantly. Women may seek hormone therapy at this time for menopause-associated symptoms such as hot flashes. Estrogen is not only important for reproductive function but also plays an important role in other tissues such as bone, the cardiovascular system and the central nervous system.

Several lines of evidence suggest that estrogen reduces the occurrence of strokes or improves outcomes from stroke. Some of this evidence comes from studies reporting sex and age differences in the incidence and outcomes of stroke, where younger women are less likely to suffer strokes than males. The second line of evidence is from the preclinical literature where levels of estrogen are manipulated and these studies have shown that estrogen treatment in young females is neuroprotective for stroke, while estrogen treatment in older females is controversial.

This chapter will consider the clinical and experimental evidence for the role of estrogen in hemorrhagic and ischemic stroke.

HEMORRHAGIC STROKE

Hemorrhagic strokes can be of two types: an aneurysm, which is the ballooning of a weakened region in the blood vessel wall that ruptures, causes bleeding and eventually results in a clot in the brain. This type of bleeding usually occurs between the arachnoid and pial coverings of the brain and is also called a subarachnoid hemorrhage (SAH). The second type is an arteriovenous malformation (AVM), where arteries and veins form abnormal connections, bypassing the capillary system and brain tissue. Such abnormal vessels are more susceptible to rupture and cause hematomas.

Gender differences have been observed in hemorrhagic strokes. More women are likely to suffer from SAH as compared to men (2:1) and female survivors are more likely to be older than male survivors. Women were also more likely to have internal carotid artery aneurysms and harbor multiple aneurysms. However, despite the younger age of SAH in men, their overall mortality and neurological outcomes are not different from females (Kongable et al., 1996). A more recent study, however, of a Dutch population, showed that, while overall mortality due to SAH has decreased, this decrease was only significant for men (Nieuwkamp et al., 2013). Seasonal variations in SAH were also more common in women than men (Ishihara et al., 2013). In younger populations where SAH is infrequent and outcomes are generally good, women were found to have a higher incidence of multiplicity and interoperative ruptures as compared to men (Park et al., 2008).

Smoking, hypertension and alcohol consumption are known risk factors for SAH. A prospective population-based cohort study found that there were no sex differences in the risk for SAH with respect to hypertension or alcohol consumption; however, cigarette smoking was a greater risk factor in women as compared to men (Lindekleiv et al., 2011; Okamoto et al., 2005).

Studies on ethnic populations show that Mexican Americans have a greater risk for SAH as compared to non-Hispanic whites. This study also confirmed the greater incidence of SAH in women as compared to men, irrespective of ethnicity (Eden et al., 2008). Overall, female sex appears to increase the incidence of hemorrhagic stroke.

In experimental studies, an intracerebral hemorrhage (ICH) is modeled in one of two ways. One procedure consists of taking blood from the animal and infusing it in the brain (autologous infusion), which mimics the blood clot seen in hemorrhagic stroke. A second procedure consists of injecting collagenase into brain vessels to weaken the vessel wall, which causes the vessel to rupture and release blood into the brain. Both methods result in similar hematomas; however, the collagenase model causes a more severe injury as measured by increased neuronal loss, increased damage to the blood–brain barrier and greater loss of motor behavior (MacLellan et al., 2008). In an animal model of SAH, males had greater mortality as compared to females, and male sex was a determining factor for greater bleeding and early injury (Friedrich et al., 2013). This evidence suggested that ovarian hormones such as estrogen might be protective for hemorrhagic stroke in females, which was directly tested in several studies.

In female rats, estrogen treatment had no effect on clot size but reduced secondary damage to the striatum and overlying cortex, and reduced mortality (Yang et al., 2001). Using a collagenase model of ICH, estradiol treatment, at various doses, to male rats has been shown to reduce striatal lesions, and a moderate dose of estradiol reduced intracranial hemorrhagic blood volume (Auriat et al., 2005). Furthermore, in an autologous blood infusion model, estradiol reduced vasospasm in male rats (Lin et al., 2006), using a mechanism involving reduction of inducible nitric oxide synthetase (Shih et al., 2006).

Estrogen treatment given after hemorrhage, however, has produced mixed results. For example, estradiol treatment to males after hemorrhage did not improve lesion volume or behavioral recovery (Nguyen et al., 2008), indicating that pretreatment with this steroid is more effective. However, daily injections of 17b-estradiol reduced cell death in SAH model in association with activation of cell survival transducers such as pAKT (Kao et al., 2013). Similarly, posthemorrhage treatment with 2-methoxyestradiol (2ME2), a naturally occurring metabolite of estradiol, reduced mortality in male rats, improved neurologic function and reduced cerebral vasospasm (Yan et al., 2006). Specifically, 2ME2 inhibition of hypoxia-inducible factor may be related to brain protection in a subarachnoid hemorrhagic model (Wu et al., 2013). SAH has been shown to increase ER-alpha but not ER-beta in the basilar artery, and ER-alpha dependent pathways are needed for relief of cerebral vasospasm (Shih et al., 2008).

Conclusion: Overall, there appears to be dichotomy between the clinical data on hemorrhagic stroke and the experimental data. Epidemiologically, the female to male mortality ratios for SAH are worse for young and older women, but there are no sex differences in data for intracerebral hemorrhage mortality. Experimental data indicate that estrogen is protective for hemorrhagic stroke outcomes, yet young women, who presumably have higher circulating levels of estrogen, have no advantage in SAH mortality. This discrepancy suggests that, while estrogen is neuroprotective, other endocrine compounds may alter outcomes for this subtype of stroke.

ISCHEMIC STROKE

Ischemic strokes (IS) are more prevalent than hemorrhagic strokes and sex differences and hormone effects have been well-studied in this type of stroke. While IS occur in both males and females, females have a higher incidence of ischemic stroke, display more nonclassical stroke symptoms and tend to have worse outcomes from stroke (Table 9.1). For example, stroke is the fourth leading cause of death overall; however, stroke is the third leading cause of death in women, and the fifth leading cause of death in men (National Center for Health Statistics, 2011). Furthermore, since women live longer than men, it is projected that stroke-related disability and institutionalization is likely to affect women more than men (Lai et al., 2005). Most ischemic strokes occur in the elderly and among this elderly demographic, women are more likely to get a stroke (Petrea et al., 2009). Women account for 60% of stroke-related deaths (Lloyd-Jones et al., 2010), even after normalization for age. The 5-year stroke recurrence is disproportionately higher in females (20%) as compared to males (10%) in the 45–64 age range (Roger et al., 2011). A Canadian stroke registry study reported that 10% of women stroke patients were discharged to long-term care as compared to 5% of men (Kapral et al., 2005), despite the observation that stroke size tends not to be different in males and females (Silva et al., 2010). In the Danish National Registry analysis, women were reported to have more severe strokes than men although they exhibited a survival advantage compared to men, especially at advanced ages (Olsen and Andersen, 2010).

Sex differences in stroke outcome are also well recognized in preclinical models. Females have a smaller infarct volume and better cerebral blood flow than age-matched males both in normal (Alkayed et al., 1998) and diabetic (Toung et al., 2000) animals. However, although female mice sustain a much smaller infarct as compared to age-matched males (Manwani et al., 2011), aged females showed significantly more mortality and poorer stroke outcomes as compared to older males. The female advantage seen in young age demographics as well as animal models has led to the hypothesis that estrogen, the major ovarian hormone, may be protective for stroke. This is supported by recent evidence that the menopause transition is a period of greater stroke risk in women (Towfighi et al., 2007).

Role of Estrogen in Ischemic Stroke

Two types of populations shed light on the issue of hormones and stroke risk: the first is young women who use oral contraceptives (OC) and the second consists of postmenopausal women who are prescribed hormone therapy after menopause. With respect to the first category, data on the effect of oral contraceptives for stroke risk in younger women is confounded by the fact that stroke incidence is relatively small in this group. A meta-analysis of 16 studies on stroke risk and OC use concluded that while there was an overall increased risk for stroke among OC users, the absolute risk was considered small since stroke is relatively rare in this population (Gillum et al., 2000). Hypertension, diabetes mellitus, hypercholesterolemia and smoking were much larger contributing factors for stroke in this population as compared to OC use (Balci et al., 2011). Cumulative data on OC and stroke risk is further complicated by the fact that OC formulations have changed over the last 50 years. For example, low dose (<50 micrograms) OC did not affect the odds ratio but high doses of OC increased stroke risk fivefold (World Health Organization Study, 1996). Additionally moderate estrogen doses with third-generation progestins significantly reduced stroke (Lidegaard, 1998).

Hormone use and stroke incidence in postmenopausal women present a more complicated picture. An early case-control study reported no increased risk for stroke in postmenopausal women who took hormone therapy (Petitti et al., 1998). In a multicenter case-controlled study, increased lifetime exposure to estrogen was associated with a lower risk of stroke, but interestingly a lower age at menarche increased the odds of stroke (de Lecinana et al., 2007). The Women's Health Initiate (WHI) study indicated that hormone use increased stroke risk. In this randomized, double blind, placebo-controlled multicenter trial, conjugated equine estrogens (CEE) (Hendrix et al., 2006) and CEE + progestins (Wassertheil-Smoller et al., 2003) increased the risk for stroke. This study recruited women from ages 50 to 85 years of age, and subsequent subgroup analyses indicated that most of the stroke risk due to hormone therapy was seen in the older age groups. In the CEE trial, increased risk for stroke was statistically significant for the 60–69-yr-old group but not in the 50–59-yr-old group (Hendrix et al., 2006). In an observational analysis of postmenopausal women in the Nurse's Health Study, estrogen and estrogen + progestin use increased the risk of stroke irrespective of the age of the user or time since menopause (Grodstein et al., 2008). However, the observational arm of the WHI study showed no increased risk for stroke in the CEE or CEE + progestin arm (Prentice et al., 2005, 2006). A possible factor in the discrepancy between the WHI trial and the WHI observational study was that the initiation of hormones was much earlier in the latter study. It should also be noted that users randomized to the HT arm differ in important physiological traits from women who seek HT independently. For example, an observational trial (SHOW study) reported no difference in stroke risk in HT users; however, HT users were more likely to have normal blood pressure and low body mass as compared to nonusers in this study (Bushnell, 2009), which was not the case in the WHI study, where hypertension incidence was similar in CEE users and nonusers (Hendrix et al., 2006). A similar interaction between HT and hypertension was seen in the Danish Nurses study, where the risk for stroke was most prominent among hypertensive women who used hormone therapy (Lokkegaard et al., 2003), while normotensive women who used hormone therapy were no different from controls.

Animal Models of Stroke and Estrogen Treatment

Although experimental stroke studies do not assess risk, they have been instrumental in determining the effect of estrogen on stroke severity.

Ischemic Stroke Models

Several models of ischemic stroke are used and most of these involve obstruction of the middle cerebral artery (MCA), which supplies blood to the striatum and overlying cortex (reviewed in McRae and Carswell, 2006). Permanent ischemic models (where circulation is never re-established) include inserting a transluminal suture, electrocautery or ligation of the MCA. Sustained (but not permanent) occlusions can be obtained by injections of the vasoconstrictor endothelin-1, while transient ischemic models can be performed by inserting and then removing the intraluminal suture. The model that most closely resembles a human ischemic stroke is the embolic model where small blood clots or artificial spheres are injected into the vessel to cause blockage of blood flow. This latter model results in variable infarctions and is therefore infrequently used. Irrespective of the method, disruption to the MCA results in an infarct involving the striatum and overlying cortex. Other models such as the carotid artery ligation model result in hippocampal damage.

Similar to human stroke, experimental ischemic stroke models have shown that female sex appears to be protective, such that females have a smaller infarct volume and better cerebral blood flow than age-matched males (Alkayed et al., 1998). However, this female advantage is lost when the animals are bilaterally ovariectomized. Interestingly, estrogen replacement to ovariectomized normoglycemic females is neuroprotective, but estrogen replacement to diabetic mice (db/db) did not reduce infarct size or regulate apoptotic genes (Zhang et al., 2004). Stroke injury to females in proestrus (high estrogen levels) results in smaller infarcts than those in metestrus (low estrogen state) and the extent of ischemic damage was inversely related to circulating levels of estrogen (Liao et al., 2001). Bilateral ovariectomy worsens infarct volume and longer periods of estrogen deprivation (1 week versus 4 weeks of ovariectomy) further increase the size of the infarct (Fukuda et al., 2000).

One of the first studies to show that estrogen had a neuroprotective effect on ischemic stroke injury came from the work of J. Simpkins, who reported that replacement with either 17β-estradiol or its inactive stereoisomer 17α-estradiol both reduced infarct size and attenuated mortality in a transient MCA occlusion model (Simpkins et al., 1997). Subsequently these studies have been expanded to show that exogenous estrogen replacement is neuroprotective when given prior (Dubal et al., 1998; Rusa et al., 1999) or subsequent to the injury (Liu et al., 2007; Yang et al., 2003), and is also effective in males (Toung et al., 1998). Similarly, conjugate equine estrogens (CEE), which is the hormone preparation used by women, is also neuroprotective in animal stroke (McCullough et al., 2001).

ESTROGEN EFFECTS ON TRAUMATIC BRAIN INJURY

However, estrogen replacement is not neuroprotective in all ischemic stroke models. Estrogen's neuroprotective effects are more consistently seen in cortical areas rather than subcortical regions, although in a less severe stroke (30 min ischemia versus 60 min ischemia), estrogen may also protect from subcortical damage (Fan et al., 2003). Estrogen replacement to the Wistar-Kyoto rat strain reportedly increases infarct volume (Carswell et al., 2004). In view of the observation that hypertension in women may modulate the effect of estrogen on stroke risk, it is worth noting that estrogen replacement had no protective effect on infarct size in the stroke prone spontaneously hypertensive rat (Carswell et al., 2000). In a four-vessel occlusion model, where both the vertebral arteries are cauterized and carotid arteries reversibly compressed, larger hippocampal injury is seen in intact and estrogen-replaced ovariectomized females as compared to ovariectomized females (Harukuni et al., 2001), although other laboratories have shown that estrogen is neuroprotective in this model (Brann et al., 2007). In an instance of severe ischemic injury, where cerebral vessels (single middle cerebral artery [MCA] and both common carotids) were occluded for 3 h, there were no gender differences in infarct size and no reduction of the infarct due to intravenous or subcutaneous 17b-estradiol (Vergouwen et al., 2000). Based on their studies where estrogen fails to protect against stroke in the ovariectomized spontaneously hypertensive stroke prone rat (Carswell et al., 2004), and actually increases damage in the Wistar-Kyoto rats, Lister Hooded and Sprague-Dawley rats (Bingham et al., 2005; Gordon et al., 2005), Macrae and Carswell (2006) have suggested that the neuroprotective effect of estrogens may be less effective in permanent ischemic models.

Since ischemic stroke occurs primarily in an older population, the role of estrogens in older animals is vital to predicting its role as a neuroprotectant. Relatively few studies have assessed older females in experimental stroke models. Stroke damage is worse in older females as compared to younger females (Liu et al., 2009b; Selvamani and Sohrabji, 2010a; Takaba et al., 2004). However, studies differ in their conclusions regarding estrogen treatment in older females. Estrogen replacement is neuroprotective in both young and middle-aged females in the MCAo suture stroke model (Alkayed et al., 2000; Dubal and Wise, 2001), although estrogens failed to attenuate hippocampal cell death in a bilateral carotid artery occlusion model in middle-aged gerbils (De Butte-Smith et al., 2007). In middle-aged female rats, characterized as reproductively senescent by daily vaginal smears and with virtually undetectable levels of estrogen (constant diestrus), estrogen treatment increased infarct volume and worsened sensory motor performance in an endothelin-1 vasoconstriction model, although estrogen treatment to multiparous young females was neuroprotective in this model

(Selvamani and Sohrabji, 2010a,b). In a study where middle-aged female rats were treated with estradiol for 9 months, estrogen treatment was also found not to be neuroprotective in an ischemic stroke model (Leon et al., 2012). Hence, estrogen treatment in older females is less likely to be neuroprotective, and this may be linked to the severity of the stroke model or to differences in the definition of middle age. Middle-aged animals defined as reproductively senescent have experienced a period of hypoestrogenicity and this may make their response to subsequent estrogen treatment less favorable. In this reproductively senescent group, estrogen treatment failed to reduce infarct volume when subject to the MCAo suture stroke model, suggesting that a period of hypo-estrogenicity may be more critical than the type of stroke model. This is further supported by a study by Wise and colleagues, in which ovariectomized females were replaced with estrogen either immediately or 10 weeks later. Immediate estrogen replacement reduced infarct volume and attenuated the inflammatory response, while estrogen replacement after a prolonged period of hypoestrogenicity was ineffective (Suzuki et al., 2007).

THE CONCEPT OF TIMING

In 1997, a mathematical model of several epidemiological studies concluded that the benefit from cardiovascular disease provided by hormone therapy far outweighed the potential risks and recommended a broader use of hormone therapy in postmenopausal women (Col et al., 1997). Ironically, a decade later, the Women's Health Initiative (WHI) study yielded just the opposite recommendation, namely that hormone therapy should be given in the smallest dose possible for the shortest amount of time and should not be prescribed for cardiovascular protection (FDA statement posted 8/13/2002). The current dilemma, therefore, is reconciling the overwhelming beneficial effects conferred by estrogen in animal models of stroke with the findings of clinical trials. One of the main hypotheses to emerge from these efforts is the *timing* hypothesis, which proposes that hormone therapy is more benign for stroke when taken by younger women or during the perimenopausal or early postmenopausal period, but deleterious when taken by women significantly past menopause (Barrett-Connor, 2007; Choi et al., 2011).

Subanalyses of the WHI data, described earlier, showing that younger women did not have deleterious outcomes from HT use support the timing hypothesis. Other evidence for the timing hypothesis also comes from a prospective study of Swedish women, where stroke risk was decreased in women who initiated hormone treatment prior to menopause (Li et al., 2006). In a population-based nested case-control study of 50–69-yr-old women, HT did not significantly elevate ischemic stroke (Arana et al., 2006), further supporting the idea that HT at ages closer to menopause may be harmless for stroke. Coronary artery calcification, a surrogate marker of cardiac disease, was reduced by estrogen in the youngest cohort of WHI (50–59 years) (Manson et al., 2007), signifying that estrogen's effects can be modulated by the age of the user. Finally, a study of non-HT users found that stroke-related mortality in women 65 and older was higher in women with higher levels of endogenous estrogen (Maggio et al., 2009), implying that elevated levels of hormones in late life, whether exogenous or endogenous, may exert a deleterious effect on stroke.

While the timing hypothesis reconciles the disparate findings of hormone therapy, testing this hypothesis is hampered by the lack of an operational definition of timing. Most authors agree that it constitutes a certain period of time during the perimenopause; however, a more specific biochemical signature would be critical for designing testable studies and for eventually tailoring medical practice. Animal studies provide the best avenue for developing markers of the critical period for hormone effectiveness. One hypothesis is that estrogen and hormone therapy is modulated by the overall endocrine status of the individual (Sohrabji et al., 2013). Specifically, the gradual loss of effectiveness of estrogen as a neuroprotectant is due to age-dependent loss of other endocrine agents that complement or collaborate with estrogen.

SEX DIFFERENCES IN ISCHEMIC STROKE THERAPIES

Besides sex differences in incidence and mortality in stroke, new data shows that therapies for stroke may also be sex specific. Much of this evidence comes from the preclinical literature where new neuroprotective compounds are being tested. For example, the anti-inflammatory drug, minocycline, which is very effective in reducing infarct volume in male rats, is completely ineffective in females (Li and McCullough, 2009). Another class of compounds called microRNA also showed sex-specific effects. Inhibitors to the miRNA Let-7f decreased brain infarct volume in intact females but had no effect in males or ovariectomized females (Selvamani et al., 2012). These data suggest that therapeutic compounds will interact with the endocrine environment of the stroke patient to modify outcomes. In females, caspase activation, which mediates cell death, is enhanced after stroke, and inhibiting this pathway benefits females but not males in both neonatal populations (Renolleau et al., 2007) and adult animals (Liu et al., 2009a). While compounds may be effective in one sex and not in the other, there is also danger that it may be effective in one sex and deleterious for the other. One such example is the PJ34 compound, which is an inhibitor to the DNA repair enzyme PARP. In males, the PARP inhibitor significantly reduced the extent of brain infarction by almost two-thirds, while in females the extent of brain tissue damage was increased almost fivefold (Yuan et al., 2009). These preclinical studies suggest that the hormonal environment of the stroke patient should be a prime consideration in the design of stroke therapies.

SEX-SPECIFIC GUIDELINES FOR STROKE

Based on the accumulated evidence of sex differences in the incidence, mortality and outcome for stroke, the American Heart Association/American Stroke Association recently convened a group of physicians/scientists to review the evidence for sex differences in stroke and the evidence for sex-specific risk factors for stroke (Bushnell et al., 2014). This writing group found that, while there were risk factors with similar prevalence in males and females, specific risk factors were found to be stronger or more prevalent in females than males. Migraine, atrial fibrillation, Type 2 diabetes, and hypertension are all stroke risk factors that were more likely to impact women than men. Finally, they identified a group of risk factors that only occurred in women, which included pregnancy and associated events such as gestational diabetes, pre-eclampsia, as well as use of hormones either as oral contraceptives or postmenopausal estrogen therapy. Their conclusions were that health care providers needed a sex-specific stroke score to provide better care for female patients.

References

Alkayed, N.J., Harukuni, I., Kimes, A.S., London, E.D., Traystman, R.J., Hurn, P.D., et al., 1998. Gender-linked brain injury in experimental stroke. Stroke. 29, 159–165, discussion 166.

Alkayed, N.J., Murphy, S.J., Traystman, R.J., Hurn, P.D., Miller, V.M., 2000. Neuroprotective effects of female gonadal steroids in reproductively senescent female rats. Stroke. 31, 161–168.

Arana, A., Varas, C., Gonzalez-Perez, A., Gutierrez, L., Bjerrum, L., Garcia Rodriguez, L. A., et al., 2006. Hormone therapy and cerebrovascular events: a population-based nested case-control study. Menopause. 13, 730–736.

Auriat, A., Plahta, W.C., McGie, S.C., Yan, R., Colbourne, F., 2005. 17beta-Estradiol pretreatment reduces bleeding and brain injury after intracerebral hemorrhagic stroke in male rats. J. Cereb. Blood Flow Metab. 25, 247–256.

Azcoitia, I., Santos-Galindo, M., Arevalo, M.A., Garcia-Segura, L.M., 2010. Role of astroglia in the neuroplastic and neuroprotective actions of estradiol. Eur. J. Neurosci. 32, 1995–2002.

Balci, K., Utku, U., Asil, T., Celik, Y., 2011. Ischemic stroke in young adults: risk factors, subtypes, and prognosis. Neurologist. 17, 16–20.

Barrett-Connor, E., 2007. Hormones and heart disease in women: the timing hypothesis. Am. J. Epidemiol. 166, 506–510.

Bingham, D., Macrae, I.M., Carswell, H.V., 2005. Detrimental effects of 17beta-oestradiol after permanent middle cerebral artery occlusion. J. Cereb. Blood Flow Metab. 25, 414–420.

Brann, D.W., Dhandapani, K., Wakade, C., Mahesh, V.B., Khan, M.M., 2007. Neurotrophic and neuroprotective actions of estrogen: basic mechanisms and clinical implications. Steroids. 72, 381–405.

Bushnell, C., 2009. Stroke Hormones and Outcomes in Women (SHOW) study: is the "healthy-user effect" valid for women after stroke? Womens Health (Lond Engl). 5, 485–496.

Bushnell, C., McCullough, L.D., Awad, I.A., Chireau, M.V., Fedder, W.N., Furie, K.L., et al., 2014. Guidelines for the prevention of stroke in women: a statement for healthcare professionals from the American Heart Association/American Stroke Association. Stroke. 45, 1545–1588.

Carswell, H.V., Dominiczak, A.F., Macrae, I.M., 2000. Estrogen status affects sensitivity to focal cerebral ischemia in stroke-prone spontaneously hypertensive rats. Am. J. Physiol. Heart Circ. Physiol. 278, H290–H294.

Carswell, H.V., Bingham, D., Wallace, K., Nilsen, M., Graham, D.I., Dominiczak, A.F., et al., 2004. Differential effects of 17beta-estradiol upon stroke damage in stroke prone and normotensive rats. J. Cereb. Blood Flow Metab. 24, 298–304.

Carswell, H.V., Dominiczak, A.F., Garcia-Segura, L.M., Harada, N., Hutchison, J.B., Macrae, I.M., et al., 2005. Brain aromatase expression after experimental stroke: topography and time course. J. Steroid Biochem. Mol. Biol. 96, 89–91.

Choi, S.D., Steinberg, E.M., Lee, H.H., Naftolin, F., 2011. The Timing Hypothesis remains a valid explanation of differential cardioprotective effects of menopausal hormone treatment. Menopause. 18, 230–236.

Col, N.F., Eckman, M.H., Karas, R.H., Pauker, S.G., Goldberg, R.J., Ross, E.M., et al., 1997. Patient-specific decisions about hormone replacement therapy in postmenopausal women. JAMA. 277, 1140–1147.

De Butte-Smith, M., Nguyen, A.P., Zukin, R.S., Etgen, A.M., Colbourne, F., 2007. Failure of estradiol to ameliorate global ischemia-induced CA1 sector injury in middle-aged female gerbils. Brain Res. 1153, 214–220.

de Lecinana, M.A., Egido, J.A., Fernandez, C., Martinez-Vila, E., Santos, S., Morales, A., et al., 2007. Risk of ischemic stroke and lifetime estrogen exposure. Neurology. 68, 33–38.

Dubal, D., Kashon, M., Pettigrew, L., Ren, J., Finklestein, S., Rau, S., et al., 1998. Estradiol protects against ischemic injury. J. Cereb. Blood Flow Metab. 18, 1253–1258.

Dubal, D.B., Wise, P.M., 2001. Neuroprotective effects of estradiol in middle-aged female rats. Endocrinology. 142, 43–48.

Eden, S.V., Meurer, W.J., Sánchez, B.N., Lisabeth, L.D., Smith, M.A., Brown, D.L., et al., 2008. Gender and ethnic differences in subarachnoid hemorrhage. Neurology. 71, 731–735.

Fan, T., Yang, S.H., Johnson, E., Osteen, B., Hayes, R., Day, A.L., et al., 2003. 17beta-Estradiol extends ischemic thresholds and exerts neuroprotective effects in cerebral subcortex against transient focal cerebral ischemia in rats. Brain Res. 993, 10–17.

Friedrich, V., Bederson, J.B., Sehba, F.A., 2013. Gender influences the initial impact of subarachnoid hemorrhage: an experimental investigation. PLoS One. 8 (11), e80101.

Fukuda, K., Yao, H., Ibayashi, S., Nakahara, T., Uchimura, H., Fujishima, M., et al., 2000. Ovariectomy exacerbates and estrogen replacement attenuates photothrombotic focal ischemic brain injury in rats. Stroke. 31, 155–160.

Gillum, L.A., Mamidipudi, S.K., Johnston, S.C., 2000. Ischemic stroke risk with oral contraceptives: a meta-analysis. JAMA. 284, 72–78.

Gordon, K.B., Macrae, I.M., Carswell, H.V., 2005. Effects of 17beta-oestradiol on cerebral ischaemic damage and lipid peroxidation. Brain Res. 1036, 155–162.

Grodstein, F., Manson, J.E., Stampfer, M.J., Rexrode, K., 2008. Postmenopausal hormone therapy and stroke: role of time since menopause and age at initiation of hormone therapy. Arch. Intern. Med. 168, 861–866.

Harukuni, I., Hurn, P.D., Crain, B.J., 2001. Deleterious effect of beta-estradiol in a rat model of transient forebrain ischemia. Brain Res. 900, 137–142.

Hendrix, S.L., Wassertheil-Smoller, S., Johnson, K.C., Howard, B.V., Kooperberg, C., Rossouw, J.E., et al., 2006. Effects of conjugated equine estrogen on stroke in the Women's Health Initiative. Circulation. 113, 2425–2434.

Ishihara, H., Kunitsugu, I., Nomura, S., Koizumi, H., Yoneda, H., Shirao, S., et al., 2013. Seasonal variation in the incidence of aneurysmal subarachnoid hemorrhage associated with age and gender: 20-year results from the Yamaguchi cerebral aneurysm registry. Neuroepidemiology. 41, 7–12.

Kao, C.H., Chang, C.Z., Su, Y.F., Tsai, Y.J., Chang, K.P., Lin, T.K., et al., 2013. 17β-Estradiol attenuates secondary injury through activation of Akt signaling via estrogen receptor alpha in rat brain following subarachnoid hemorrhage. J. Surg. Res. 183, e23–e30.

Kapral, M.K., Fang, J., Hill, M.D., Silver, F., Richards, J., Jaigobin, C., et al., 2005. Sex differences in stroke care and outcomes: results from the Registry of the Canadian Stroke Network. Stroke. 36, 809–814.

Kongable, G.L., Lanzino, G., Germanson, T.P., Truskowski, L.L., Alves, W.M., Torner, J.C., et al., 1996. Gender-related differences in aneurysmal subarachnoid hemorrhage. J. Neurosurg. 84, 43–48.

Lai, S.M., Duncan, P.W., Dew, P., Keighley, J., 2005. Sex differences in stroke recovery. Prev. Chronic. Dis. 2, A13.

Leon, R.L., Li, X., Huber, J.D., Rosen, C.L., 2012. Worsened outcome from middle cerebral artery occlusion in aged rats receiving 17b-estradiol. Endocrinology. 153, 3386–3393.

Li, C., Engstrom, G., Hedblad, B., Berglund, G., Janzon, L., 2006. Risk of stroke and hormone replacement therapy. A prospective cohort study. Maturitas. 54, 11–18.

Li, J., McCullough, L.D., 2009. Sex differences in minocycline-induced neuroprotection after experimental stroke. J. Cereb. Blood Flow Metab. 29, 670–674.

Liao, S., Chen, W., Kuo, J., Chen, C., 2001. Association of serum estrogen level and ischemic neuroprotection in female rats. Neurosci. Lett. 297, 159–162.

Lidegaard, O., 1998. Thrombotic diseases in young women and the influence of oral contraceptives. Am. J. Obstet. Gynecol. 179, S62–S67.

Lin, C.L., Shih, H.C., Dumont, A.S., Kassell, N.F., Lieu, A.S., Su, Y.F., et al., 2006. The effect of 17beta-estradiol in attenuating experimental subarachnoid hemorrhage-induced cerebral vasospasm. J. Neurosurg. 104, 298–304.

Lindekleiv, H., Sandvei, M.S., Njølstad, I., Løchen, M.L., Romundstad, P.R., Vatten, L., et al., 2011. Sex differences in risk factors for aneurysmal subarachnoid hemorrhage: a cohort study. Neurology. 76, 637–643.

Liu, F., Li, Z., Li, J., Siegel, C., Yuan, R., McCullough, L.D., et al., 2009a. Sex differences in caspase activation after experimental stroke. Stroke. 40, 1842–1848.

Liu, F., Yuan, R., Benashski, S.E., McCullough, L.D., 2009b. Changes in experimental stroke outcome across the life span. J. Cereb. Blood Flow Metab. 29, 792–802.

Liu, R., Wang, X., Liu, Q., Yang, S.H., Simpkins, J.W., 2007. Dose dependence and therapeutic window for the neuroprotective effects of 17beta-estradiol when administered after cerebral ischemia. Neurosci. Lett. 415, 237–241.

Lloyd-Jones, D., Adams, R.J., Brown, T.M., Carnethon, M., Dai, S., De Simone, G., et al., 2010. Heart disease and stroke statistics—2010 update: a report from the American Heart Association. Circulation. 121, e46–e215.

Lokkegaard, E., Jovanovic, Z., Heitmann, B.L., Keiding, N., Ottesen, B., Hundrup, Y.A., et al., 2003. Increased risk of stroke in hypertensive women using hormone therapy: analyses based on the Danish Nurse Study. Arch. Neurol. 60, 1379–1384.

MacLellan, C.L., Silasi, G., Poon, C.C., Edmundson, C.L., Buist, R., Peeling, J., et al., 2008. Intracerebral hemorrhage models in rat: comparing collagenase to blood infusion. J. Cereb. Blood Flow Metab. 28, 516–525.

MacRae, I.M., Carswell, H.V., 2006. Oestrogen and stroke: the potential for harm as well as benefit. Biochem. Soc. Trans. 34, 1362–1365.

Maggio, M., Ceda, G.P., Lauretani, F., Bandinelli, S., Ruggiero, C., Guralnik, J.M., et al., 2009. Relationship between higher estradiol levels and 9-year mortality in older women: the Invecchiare in Chianti study. J. Am. Geriatr. Soc. 57, 1810–1815.

Manson, J.E., Allison, M.A., Rossouw, J.E., Carr, J.J., Langer, R.D., Hsia, J., et al., 2007. Estrogen therapy and coronary-artery calcification. N. Engl. J. Med. 356, 2591–2602.

Manwani, B., Liu, F., Xu, Y., Persky, R., Li, J., McCullough, L.D., et al., 2011. Functional recovery in aging mice after experimental stroke. Brain Behav. Immun. 25, 1689–1700.

McCullough, L.D., Alkayed, N.J., Traystman, R.J., Williams, M.J., Hurn, P.D., 2001. Postischemic estrogen reduces hypoperfusion and secondary ischemia after experimental stroke. Stroke. 32, 796–802.

National Center for Health Statistics. Health, United States, 2011. Hyattsville, MD: US Department of Health and Human Services, Centers for Disease Control and Prevention, National Center for Health Statistics; 2012. Available from: <http://www.cdc.gov/nchs/hus/contents2011.htm#031>.

Nguyen, A.P., Arvanitidis, A.P., Colbourne, F., 2008. Failure of estradiol to improve spontaneous or rehabilitation-facilitated recovery after hemorrhagic stroke in rats. Brain Res. 1193, 109–119.

Nieuwkamp, D.J., Vaartjes, I., Algra, A., Bots, M.L., Rinkel, G.J., 2013. Age- and gender-specific time trend in risk of death of patients admitted with aneurysmal subarachnoid hemorrhage in the Netherlands. Int. J. Stroke. Suppl. A100, 90–94.

Okamoto, K., Horisawa, R., Ohno, Y., 2005. The relationships of gender, cigarette smoking, and hypertension with the risk of aneurysmal subarachnoid hemorrhage: a case-control study in Nagoya, Japan. Ann. Epidemiol. 15, 744–748.

Olsen, T.S., Andersen, K.K., 2010. Female survival advantage relates to male inferiority rather than female superiority: a hypothesis based on the impact of age and stroke severity on 1-week to 1-year case fatality in 40,155 men and women. Gend. Med. 7, 284–295.

Park, S.K., Kim, J.M., Kim, J.H., Cheong, J.H., Bak, K.H., Kim, C.H., et al., 2008. Aneurysmal subarachnoid hemorrhage in young adults: a gender comparison study. J. Clin. Neurosci. 15, 389–392.

Petitti, D.B., Sidney, S., Quesenberry Jr., C.P., Bernstein, A., 1998. Ischemic stroke and use of estrogen and estrogen/progestogen as hormone replacement therapy. Stroke. 29, 23–28.

Petrea, R.E., Beiser, A.S., Seshadri, S., Kelly-Hayes, M., Kase, C.S., Wolf, P.A., et al., 2009. Gender differences in stroke incidence and poststroke disability in the Framingham heart study. Stroke. 40, 1032–1037.

Prentice, R.L., Langer, R., Stefanick, M.L., Howard, B.V., Pettinger, M., Anderson, G., et al., 2005. Combined postmenopausal hormone therapy and cardiovascular disease: toward resolving the discrepancy between observational studies and the Women's Health Initiative clinical trial. Am. J. Epidemiol. 162, 404–414.

Prentice, R.L., Langer, R.D., Stefanick, M.L., Howard, B.V., Pettinger, M., Anderson, G.L., et al., 2006. Combined analysis of Women's Health Initiative observational and clinical trial data on postmenopausal hormone treatment and cardiovascular disease. Am. J. Epidemiol. 163, 589–599.

Renolleau, S., Fau, S., Goyenvalle, C., Charriaut-Marlangue, C., 2007. Sex, neuroprotection, and neonatal ischemia. Dev. Med. Child Neurol. 49, 477.

Roger, V.L., Go, A.S., Lloyd-Jones, D.M., Adams, R.J., Berry, J.D., Brown, T.M., et al., 2011. Heart disease and stroke statistics—2011 update: a report from the American Heart Association. Circulation. 123, e18–e209.

Rusa, R., Alkayed, N.J., Crain, B.J., Traystman, R.J., Kimes, A.S., London, E.D., et al., 1999. 17beta-estradiol reduces stroke injury in estrogen-deficient female animals. Stroke. 30, 1665–1670.

Selvamani, A., Sohrabji, F., 2010a. Reproductive age modulates the impact of focal ischemia on the forebrain as well as the effects of estrogen treatment in female rats. Neurobiol. Aging. 31, 1618–1628.

Selvamani, A., Sohrabji, F., 2010b. The neurotoxic effects of estrogen on ischemic stroke in older female rats is associated with age-dependent loss of insulin-like growth factor-1. J. Neurosci. 30, 6852–6861.

Selvamani, A., Sathyan, P., Miranda, R.C., Sohrabji, F., 2012. An antagomir to microRNA Let7f promotes neuroprotection in an ischemic stroke model. PLoS One. 7 (2), e32662.

Shih, H.C., Lin, C.L., Lee, T.Y., Lee, W.S., Hsu, C., 2006. 17beta-Estradiol inhibits subarachnoid hemorrhage-induced inducible nitric oxide synthase gene expression by interfering with the nuclear factor kappa B transactivation. Stroke. 37, 3025–3031.

Shih, H.C., Lin, C.L., Wu, S.C., Kwan, A.L., Hong, Y.R., Howng, S.L., et al., 2008. Upregulation of estrogen receptor alpha and mediation of 17beta-estradiol vasoprotective effects via estrogenreceptor alpha in basilar arteries in rats after experimental subarachnoid hemorrhage. J. Neurosurg. 109, 92–99.

Silva, G.S., Lima, F.O., Camargo, E.C., Smith, W.S., Lev, M.H., Harris, G.J., et al., 2010. Gender differences in outcomes after ischemic stroke: role of ischemic lesion volume and intracranial large-artery occlusion. Cerebrovasc. Dis. 30, 470–475.

Simpkins, J.W., Rajakumar, G., Zhang, Y.Q., Simpkins, C.E., Greenwald, D., Yu, C.J., et al., 1997. Estrogens may reduce mortality and ischemic damage caused by middle cerebral artery occlusion in the female rat. J. Neurosurg. 87, 724–730.

Sohrabji, F., Selvamani, A., Balden, R., 2013. Revisiting the timing hypothesis: Biomarkers that define the therapeutic window of estrogen for stroke. Horm. Behav. 63, 222–230.

Suzuki, S., Brown, C.M., Dela Cruz, C.D., Yang, E., Bridwell, D.A., Wise, P.M., et al., 2007. Timing of estrogen therapy after ovariectomy dictates the efficacy of its neuroprotective and antiinflammatory actions. Proc. Natl. Acad. Sci. U.S.A. 104, 6013–6018.

Takaba, H., Fukuda, K., Yao, H., 2004. Substrain differences, gender, and age of spontaneously hypertensive rats critically determine infarct size produced by distal middle cerebral artery occlusion. Cell. Mol. Neurobiol. 24, 589–598.

Toung, T.J., Traystman, R.J., Hurn, P.D., 1998. Estrogen-mediated neuroprotection after experimental stroke in male rats. Stroke. 29, 1666–1670.

Toung, T.K., Hurn, P.D., Traystman, R.J., Sieber, F.E., 2000. Estrogen decreases infarct size after temporary focal ischemia in a genetic model of type 1 diabetes mellitus. Stroke. 31, 2701–2706.

Towfighi, A., Saver, J.L., Engelhardt, R., Ovbiagele, B., 2007. A midlife stroke surge among women in the United States. Neurology. 69, 1898–1904.

Vergouwen, M.D., Anderson, R.E., Meyer, F.B., 2000. Gender differences and the effects of synthetic exogenous and non-synthetic estrogens in focal cerebral ischemia. Brain Res. 878, 88–97.

Wassertheil-Smoller, S., Hendrix, S.L., Limacher, M., Heiss, G., Kooperberg, C., Baird, A., et al., 2003. Effect of estrogen plus progestin on stroke in postmenopausal women: the Women's Health Initiative: a randomized trial. JAMA. 289, 2673–2684.

WHO Collaborative Study of Cardiovascular Disease and Steroid Hormone Contraception, 1996. Haemorrhagic stroke, overall stroke risk, and combined oral contraceptives: results of an international, multicentre, case-control study. The Lancet. 348, 505–510.

Wu, C., Hu, Q., Chen, J., Yan, F., Li, J., Wang, L., et al., 2013. Inhibiting HIF-1α by 2ME2 ameliorates early brain injury after experimental subarachnoid hemorrhage in rats. Biochem. Biophys. Res. Commun. 437, 469–474.

Yan, J., Chen, C., Lei, J., Yang, L., Wang, K., Liu, J., et al., 2006. 2-methoxyestradiol reduces cerebral vasospasm after 48 hours of experimental subarachnoid hemorrhage in rats. Exp. Neurol. 202, 348–356.

Yang, S.H., He, Z., Wu, S.S., He, Y.J., Cutright, J., Millard, W.J., et al., 2001. 17-beta estradiol can reduce secondary ischemic damage and mortality of subarachnoid hemorrhage. J. Cereb. Blood Flow Metab. 21, 174–181.

Yang, S.H., Liu, R., Wu, S.S., Simpkins, J.W., 2003. The use of estrogens and related compounds in the treatment of damage from cerebral ischemia. Ann. N. Y. Acad. Sci. 1007, 101–107.

Yuan, M., Siegel, C., Zeng, Z., Li, J., Liu, F., McCullough, L.D., et al., 2009. Sex differences in the response to activation of the poly (ADP-ribose) polymerase pathway after experimental stroke. Exp. Neurol. 217, 210–218.

Zhang, L., Nair, A., Krady, K., Corpe, C., Bonneau, R.H., Simpson, I.A., et al., 2004. Estrogen stimulates microglia and brain recovery from hypoxia-ischemia in normoglycemic but not diabetic female mice. J. Clin. Invest. 113, 85–95.

CHAPTER

10

Concluding Statements and Current Challenges

Kelli A. Duncan

Biology and Neuroscience and Behavior, Vassar College, Poughkeepsie, New York, USA

My goal in assembling these chapters was to provide a "one-stop" resource on the new and exciting world of estrogen-mediated neuroprotection. Furthermore, I wanted to provide a resource to younger scientists to show that there are many paths to doing this type of research and that serendipitous, or eureka, moments can still occur and change both the trajectory of your life and your research. Our understanding of estrogen and estrogen-mediated neuroprotection has progressed at a spectacular rate over the last 75 years. A simple publication search produces over 500 different manuscripts examining estrogens and brain damage alone. We now have definitive proof that estrogen actions are not limited to the development of sex differences and sexual differentiation of the brain and behavior, but can in fact have profound and long-lasting effects following damage to the brain.

A number of major concepts have emerged and were shown to be broadly applicable across vertebrates. As illustrated in this book, it was established, among other things, that:

1. Following damage to the brain a number of molecular and neuroimmunological processes are initiated, one such process being the induction of aromatase (estrogen synthase) from glial cells.
2. These estrogen-producing glial cells are involved in brain cell proliferation, differentiation, migration and survival in both the uninjured and injured brain.
3. Traumatic brain injury (TBI) is complex pathophysiology that requires a multifactorial therapeutic approach.

K.A. Duncan (Ed): *Estrogen Effects on Traumatic Brain Injury.*
DOI: http://dx.doi.org/10.1016/B978-0-12-801479-0.00010-3

4. Estrogens have a pleiotropic neuroprotective effect that uniquely allows them to be used as therapeutic agents across varying types of damage to the brain.
5. Estrogens may modulate apoptotic and neuroprotective pathways, as well as the immune response following injury to the brain, thus providing a viable treatment option against neurodegeneration following neural injury.
6. Males and females differ in their response to neural damage.

These findings have answered many of the questions concerning the role of estrogens following brain damage, but many still remain unanswered:

1. How do we develop treatment options (SERMs, STEARs, use of estrogens) that target the neuroprotective effects of estrogens without the negative effects often attributed to hormone replacement therapy in women (masculinizing effects, increased cardiovascular and stroke risk, breast cancer, and dementia)?
2. What, if any, are the interactions between other steroid hormones and estrogen and what effects do exogenous SERMs and STEARs have on their actions?
3. What effect does age have on recovery from TBI, specifically regarding estrogen mediation? The CDC states that 65% of TBIs occur among children (5–18 years) and when compared with adults, younger persons are at increased risk for TBIs with increased severity and prolonged recovery. With the maturation of the endocrine system occurring during the prepubertal and pubertal years (10–17 years old), what if any effect would additional estrogens have on physiology and neurodevelopment?
4. It has become very clear that sex/gender play a key role in how the body and brain respond to injury. Thus, it is imperative that we determine how males and females differ in terms of both neural injury and estrogen treatment following injury. These data suggest that sex-specific therapies and guidelines following injury are necessary.

   As stated here previously, our understanding of estrogen and estrogen mediation has expanded immensely over the last few decades. Who knows what will be known in just a few more years?

Index

Note: Page numbers followed by "*f*" and "*t*" refer to figures and tables, respectively.

Printed in the United States
By Bookmasters